The Glory of Vedic Mathematics:
Beating the Computer

The Glory of Vedic Mathematics:
Beating the Computer

Amb. Dr. Prakash Joshi

**MOTILAL BANARSIDASS
INTERNATIONAL
DELHI**

Delhi, 2025

ISBN : 978-93-48911-94-0 (HB)
ISBN : 978-93-48911-71-1 (PB)

Also available at :
MOTILAL BANARSIDASS INTERNATIONAL
41 U.A. Bungalow Road, (Back Lane) Jawahar Nagar, Delhi-110007
4261/3 (Basement), Ansari Road, Darya Ganj, New Delhi-110002
Shop#. 6, 241, Luz Ginza Complex, Luz Corner, Mylapore, Chennai
- 600004
12/1A, 2nd Floor, Bankim Chatterjee Street, Kolkata - 700073

Stockist : Motilal Books, Ashok Rajpath, Near Kali Mandir,
Patna-800004

Printed in India
MOTILAL BANARSIDASS INTERNATIONAL

Dedication

Dedicated to my respected father
Dr. V. M. Joshi, I.C.S. F.I.A., D.Sc.

A person of great integrity whose life was dedicated
to Mathematics. He always lives within me.

K.L. Ganju, o.c.v.c. (cdr.)
Consul General (Hy.)
Advisor to the Foreign Minister
Union of the Comoros
President: HCCD – India

PREFACE

FOR 'THE GLORY OF VEDIC MATHEMATICS: BEATING THE COMPUTER'

I am very happy to write this preface for the book on Vedic Mathematics authored by Dr. Prakash Joshi. The term Vedic Mathematics might seem somewhat strange and confusing to many people, especially those who belong to earlier generations. This is because until recent times it was generally not recognized that there was something called Vedic Mathematics with its own distinctive methodology.

The Vedas are mainly associated with metaphysics and are regarded as the very fountainhead of Hinduism as well as of our ancient culture. Of course, apart from religion they cover a number of topics ranging from philosophy, ecology, ethics, to much more mundane issues like how to outwit one's rival in love.

The Vedas are mainly associated with metaphysics and are regarded as the very fountainhead of Hinduism as well as of our ancient culture. Of course, apart from religion they cover a number of topics ranging from philosophy, ecology, ethics, to much more mundane issues like how to outwit one's rival in love.

But our ancient sages had also formulated several Sutras relating to computation. In fact, in the field of mathematics too ancient Indians had notable achievements to their credit. It is well known that in the 8th century an Arabic scholar, Al-Biruni or Abu Rehan, who had travelled to India had found that the Indians knew about zero and were proficient in making calculations on the basis of the decimal system. It is in this manner that the invention of zero became known to the Arab world from where it was passed on to Europe. It is very much revolutionized the system of computation then prevailing. The ancient Greeks and Romans were not aware of the decimal system or the use of zero and as such their methods of computations were clumsy, and they found it very hard to deal with big numbers.

But the discovery of zero made by our forefathers had not come in an isolated manner. Much before that came the Vedic Sutras indicating how many computations can be done easily, and what is more important even without the use of paper and pencil by following the methods as laid down in them.

These Sutras however were not known for a long time. The credit for bringing them to light goes to a former Shankaracharya of Puri, the Late Bharti Krishna Tirthaji, in the late 19th century.

These Sutras so to say constitute the bedrock of Vedic mathematics. In this book Dr. Prakash Joshi has explained and elucidated them, and shown in detail how various computations can be done easily and, more important, mentally by following their techniques.

The Vedic method of computation is in many ways different from what is done in traditional arithmetic. The Vedic division for example starts from the right while in the traditional way of division we start from the left; the Vedic mathematics makes extensive use of base numbers like 10, 100, 1000, etc. which is not done in the normal way of computation.

This book written by Dr. Joshi has some distinctive features to which I would like to allude briefly. It clearly brings out the meaning of various Sutras which have been translated into English. He has given a large number of problems and clearly specified which should be solved mentally and which not. While Vedic mathematics and traditional methods of computation ultimately reach the same result as has to happen, knowledge of Vedic mathematics will definitely strengthen one's knowledge of our number system and facilitate dealing with numbers in a confident and smooth manner.

Perhaps the most important theme which Dr. Joshi has highlighted in this book is the fact that ancient Indians knew about infinity. This was an altogether novel concept which evolved in modern mathematics only as late as the 18th century. In other words, it was at least 2000 years after the ancient Indian sages had grasped its significance. They understood that a sum of infinite numbers could very well be finite. It was truly an revolutionary idea. This was not comprehended in rest of the world causing much confusion for a long time.

Finally, I wish to warmly compliment Dr. Prakash Joshi for producing such a fine volume which is clearly and lucidly written. It can be studied with profit even by elderly people who will find the exercises given in it to be mentally stimulating. In the end I have no hesitation in recommending this book to students and to the general public. I take this opportunity to wish Dr. Joshi all success in his literary career.

(K.L. GANJU)

Dr. Kaulgud V. V.
M. B. B. S.

Reg. No. 36119

Timings : ● Monday to Saturday : 08.00 a.m. to 11.00 a.m.
● Evening : Mon. Wed. Fri. 05.00 p.m. to 6.30 p.m.
● SUNDAY CLOSED

Clinic : Darpan Apts., Sanghavi Nagar,
Opp. Bank of Maharashtra,
Aundh, Pune - 411 007.
Mob. : 98220 26790
Email : achyutkaulgud@gmail.com

7th July 2025

I am very happy to write these few words of commendation for Dr.Prakash Joshi's book 'The Glory of Vedic Mathematics: Beating the computer'. This text explains in a lucid manner certain techniques of computation based on Vedic sutras.

Dr.Joshi has shown how problems like 99 x 99 , 93x95 , long divisions with 9 as a divisor can be carried out mentally almost instantaneously using the techniques of Vedic maths..

While this book is meant for young students, I feel it would be useful and relevant for elderly people. If they master these techniques it will stimulate their thinking and mental activity.They will also learn to visualise and hold a series of numbers in their mind.This will improve their power of concentration.In view of all this I have no hesitation to recommend this book for those of advanced age.

(Dr. Kaulgud.V.V.)

Author's Foreword

This text "The Glory of Vedic Mathematics : Beating the Computer" makes an attempt to acquaint the reader with the basics of the subject. Though Vedic mathematics is based on Vedic Sutras going back to ancient times it was introduced to the modern world only in the last century by one Krishna Tirtha (1884-1960).

He was a Vedic scholar and he construed certain basic Vedic Sutras relating to computation in modern terms. It would be appropriate to say a few words about these Sutras right in the beginning.

They are a kind of aphorisms and extremely cryptic. To give an instance; one Sutra goes only as "Vilokanam" (by simple observation). Obviously it does not make much sense. But it goes to the credit of Tiratha ji that he showed how these Sutras despite their extreme brevity can be used to for arithmetical computations. **Some of the techniques in Vedic mathematics have no parallel in the modern way of calculation.**

Let us elaborate a little.

In our traditional system while carrying out numerical calculations we do not as such observe various numbers including their digits carefully, nor try to envisage how they relate to other numbers in proximity. But this aspect assumes great importance in Vedic Mathematics. This is 'Vilokanam' Sutra. To illustrate:

Take the number 17,576. It is a cube of a certain 2-digit number between 10 and 99. But what number is it? By conventional methods there is no way of knowing this except by factorizing the above number (17,576). But as anybody can see it will involve quite a bit of computation. It would have been far more difficult had this number been the cube of a prime number and a big one at that like 83. Its cube fbecomes 5, 71, 787. We won't have found this answer without ascertaining that none of the earlier primes like 2, 3, 5, ... 23, 29, ... etc. were its factor. Obviously this is a very laborious and time-consuming procedure.

But the beauty of Vedic mathematics lies in the fact that by merely observing the above number 5, 71, 787 carefully one can know in less than 10 seconds that it is a cube of 83. To give two more instances of the operation of 'Vilokanam' Sutra (by observation).

Take the multiplications 95×97 or 93×95. These cannot be carried out mentally. But through Vedic mathematics one can obtain the answers almost instantaneously without any computation on paper. The answers to the above are : 9215 and 8835. Similarly, take the number : 1,000,0001, is it divisible by 11 or not ? The traditional system will require a detailed calculation. But Vedic mathematics brings out how the divisibility by 11 can be determined by merely looking at the number. For example, the above number is divisible by 11. This conclusion has been reached by merely observing it and counting its zeros. No calculation whatsoever was required. Here the rule is as enunciated by a relevant Vedic Sutra 'any number with even number of zeros sandwiched between two "1"s at the extremity is always divisible by 11'. In the example cited earlier there were 6 zeros in between. So the number has to be divisible by 11.

All this highlights how the 'Vilokanam' Sutra is applied in practice. In succeeding sections of this book, it is explained how through 'Vilokanam' seemingly complex problems are solved in hardly any time.

We have in this small tract illustrated how many basic operations like addition, subtraction, multiplication, etc., can be done smoothly, elegantly and quickly with the help of Vedic Sutras. However, it needs to be observed that the Vedic mathematics and the conventional mathematics ultimately lead to the same result as has to happen.

We will now briefly explain the format of this book. Based on the Vedic Sutras we have formulated various rules and formulae; the same need to be followed while carrying out computations.

But in mathematics no rule should be followed blindly merely because it has been laid down in a book or taught by a teacher. Everything has a logical reason and same has to be understood clearly by the student. Then alone can the student progress in his study of mathematics. **This is why after mentioning any rule, formula, etc., we have always elucidated the reasoning behind it.**

We all know that to ancient Indians goes the credit of discovering zero. Zero was unknown till then and therefore other civilizations had to use numerous digits and symbols to express a big number. But in India everything was simplified with the discovery of zero and just by its iteration a number could be increased on and on with greatest ease like 100, 1,000, 10,000, 1,00,000, 10,00,000 and so on and so on. Al - Khwarizmi, a Persian mathematician, introduced to the West our 10-based positional number system around 8' th century A.D.

Our number system as we all know is based on the multiples of 10, 100, 1000, etc. It remains the same, be it Vedic mathematics or the conventional computation. This is why after giving any formula, rule, etc., we have explained how it can be understood in terms of this basic structure of our numbers. **The student should always try to keep these basic precepts at the back of his mind while applying a particular formula.**

In our opinion the main advantages of learning the Vedic system of mathematics would be as follows :

i)   The knowledge of both the Vedic system and of the normal methods of computation will deepen the student's understanding of our number system. This is like knowing how to traverse from point A to point B by a number of different routes. A person who can do so surely acquires a better knowledge of the topography of that area.

ii)   It will help him to think logically.

iii)   Vedic mathematics lays stress on mental calculations. This will make the student more alert and quick in his thought processes. **Notably most of the problems set in this book are so formulated that they can be and should be solved mentally almost instantaneously.**

iv)   The student thus will develop the ability to "see" numbers clearly in his mind while carrying out computations. This does not happen in conventional methods of computation.

v)   He will have a strong foundation in mathematics. The student will inculcate the ability to appreciate the symmetry, harmony and beauty of mathematics and its logical methodology.

vi)   A word here as to for whom this book is meant won't be inappropriate. It presupposes a basic familiarity with algebra and the facility for handling big numbers. It should prove suitable for students in the age group 11 to 15/16. Moreover, its perusal is also recommended for elderly people. If they manage to do calculations like 98^2, 97×95, 99]1111, etc., without any hesitation mentally, it will undoubtedly stimulate their mind and intellect. This should impart a definite psychological benefit to them. This is precisely the viewpoint of a well-known physician in Pune, one Dr. Kaulgud V.V. His endorsement is appended at the beginning.

While studying Vedic mathematics the student should always keep the following basic precept in mind : don't believe any formula, rule, etc., merely because it is so formulated. Keep on thinking, go to the basics until you know its rationale and the reasonining behind it. This is what is called "rigour" in mathematics.

Last but not in the least, we will like to highlight that this book brings out (Chapter 11) how the concept of "infinity" was properly understood by Vedic mathematicians, and how

they could apply it to solve practical problems. The western world took 2500 years to catch up with them.

This book is dedicated to my respected father Dr. V. M. Joshi, I.C.S., F.I.A., D.Sc. He was truly an outstanding mathematician. He was my life-long teacher and from him I imbibed the concept of beauty and "rigour" of mathematics.

My utmost thanks are due to Vandanaji of M/s Asmita Enterprises, Pune, for typing out the manuscript so elegantly. Her task was especially hard as typing out numbers, mathematical formulae, etc., poses special difficulties. But she successfully overcame them.

In the end I would like to express my gratitude to the publishers, M/s. Motilal Banarasidass International, for bringing out this text in a beautiful format in a remarkably short period of time. Special commendation is due to its proprietor Shri Abhishekji for inspiring me to undertake this task. I must also make a mention of Mrs. Poonam Taneja, Production Manager of M/s Motilal Banarsidass International, for extending to me every possible cooperation.

My best wishes to the students and teachers of Vedic mathematics.

The Glory of Vedic Mathematics: Beating the Computer

Section I
"Application of the Sutra 'Vilokanam"

We will in this section study how the Sutra 'Vilokanam (by observation) is used in computation.

Testing the divisibility of a number by 2, 3, 5, 6, 8, 9 and 11 : There are simple rules which tell us whether a particular number is divisible by 2, 3, 5, 6, 9 or 11. Let us see what they are. As laid down in the above Sutra, all what you have to do is look at the number given for above determination.

Divisibility by 2:
Observe merely the last digit of the number given. **If the last digit is divisible by 2, then the entire number is divisible by 2.** Such numbers are called even numbers. Examples: 2, 10, 14, 16, 2002, 1000, 1006, etc.

Numbers which are not divisible by 2 because the last digit is not divisible by 2: 3, 5, 11, 101, 709, 1111, 1311 … .

Numbers which are divisible by 3:
You have to merely add the digits of the number given. **If the sum is divisible by 3, the whole number is divisible by 3.**

I	II	III
Number	Sum of digits	Divisibility by 3
12	1+2=3	✓
17	1+7=8	✗
21	2+1=3	✓
101	1+0+1=2	✗
1002	1+00+2=3	✓

Examples of numbers divisible by 3:
27, 81, 99, 114, 210, 1002, 10,002, 100, 002, 1,60,200, …

The student should verify for himself the validity of this rule by carrying out the actual division and convince himself that if the sum of digits of a number is divisible by 3, the whole number is ipso facts divisible by 3.

Divisibility by 4:
Observe only the last two digits of number. If they are divisible by 4, the whole number is divisible by 4. To illustrate:

I	II	III
Number	Last two digits	Divisibility by 4
120	20	✓ as 20 is divisible by 4
133	33	✗ as 33 is not divisible by 4
204	04	✓ as 04 is divisible by 4
1080	80	✓ as 80 is divisible by 4
10011	11	✗ as 11 is not divisible by 4

Apply this rule to following numbers to determine their divisibility by 4.
A) 117, B) 191, C) 200, D) 204, E) 300, F) 911, G) 1001, H) 1100, I) 1301, J) 1572

The numbers at (A), (B), (F), (G) and (I) are not divisible by 4 as their last two digits are not divisible by 4. The numbers of (C), (D), (E), (H) and (J) are divisible by 4.

It would help the student if he convinces himself that this is indeed the case by direct computation. This should also be done in case of the rules which would follow in this section.

Divisibility by 5 :
Only those numbers which have endings as 0 or 5 are divisible by 5; others are not.

Numbers which are divisible by 5:
5, 50, 65, 115, 205, 315, 410, 915, 1025.

Numbers which not divisible by 5:
7, 13, 111, 903, 1001, 2007, 3017

The student should observe that in all the numbers given above the last digit is not divisible by 5 as it is neither 0 nor 5 as required in the rule stated above.

Numbers which are divisible by 6:
Those numbers which are divisible by 2 and 3 are divisible by 6. This means that such numbers need to satisfy rules applicable to division by 2 and 3. This requires they should be even numbers and the sum of their digits should be divisible by 3.

To illustrate:

108:    Even number, sum of its digits 9 which is divisible by 3. Hence the number is divisible by 6.

216:    Even, sum of digits 9. Divisible by 3. Thus divisible by 6.

396:    Even, sum of digits 18. Divisible by 3. Thus divisible by 6.

Thus student needs to note that we are reaching this conclusion by merely observing the number and not carrying out any calculation.

Divisibility by 8:
Here the rule is its last three digits should be divisible by 8. To illustrate :

I		II	III
	Number	Last three digits	Divisibility by 8
(A)	1000	000	✓
(B)	1050	050	✗
(C)	1064	064	✓
(D)	2040	040	✓
(E)	5248	248	✓
(F)	6456	456	✓
(G)	7013	013	✗

The students should observe that numbers of (A), (C), (D), (E) and (F) are divisible by 8 as their last three digits are divisible by 8. (B) and (G) are not divisible by 8 as their last three digits are not so divisible.

Divisibility by 9:
Here the rule is analogous to that of 3. It goes as : **the sum of the digits of the number should be divisible by 9**. To illustrate :

I	II	III
Number	Sum of the digits	Divisibility by 9
(A) 99	18	✓
(B) 199	19	✗
(C) 990	18	✓
(D) 1080	9	✓
(E) 2736	18	✓
(F) 3492	18	✓
(G) 10,000	1	✗

It is to be observed that the numbers of (A), (C), (D), (E), (F) are divisible by 9 as the sums of their digits 18, 18, 9, 18, 18 are divisible by 9. But numbers at (B) and (G) are not divisible by 9.

Divisibility by 11:
Here the rule goes as follows:
Add the digits at even places, add the digits in odd places. Take the difference between these two sums. If it is divisible by 11, the whole number is divisible by 11. To illustrate :

$$123 \rightarrow Digital\ position$$
$$\uparrow\uparrow\uparrow$$

Number →   121

Digits at odd places 1 and 1;   sum = 2 (A)

Digits at even place = 2 (B)

A-B or difference of the two = 0

This is divisible by 11. Hence 121 is divisible by 11.

$$12\ 345 \rightarrow \textit{Digital position}$$
$$\uparrow \uparrow \uparrow \uparrow \uparrow$$

Number → 11,111
Digits at odd places : 1, 1, 1   Sum = 3 (A)
Digits at even places :, 1, 1.   Sum = 2 (B)
A–B or difference of the two = 1
Not divisible by 11. Hence 11,111 is not divisible by 11.

$$1234 \rightarrow \textit{Digital position}$$
$$\uparrow \uparrow \ \uparrow \uparrow$$

Number → 1243
Digits at odd place : 1,4.   Sum = 5 (A)
Digits of even places : 2, 3.   Sum of the digits = 5 (B)
A - B or difference of the two : 0

This is divisible by 11. Hence, 1243 is divisible by 11.

This is a very simple method and can be applied very fast by simple mental calculation. Take a huge number like, 1,000,0001. The student has only to observe that the first digit is at an odd place while the last digit is at even place, 8th.

The difference of the two is 0. Hence, 1,000,0001 is divisibly by 11. Here the quotient is 909091. On the other hand, $\frac{12\ 345\ 6789}{10,000,0001}$ is not divisible by 11 as both the "1" digits are at odd places [1 and 9].

Hence, sums of even/odd digits equal : 0 and 2.
Difference = 2. Not divisible by 11.

Hence, 10,000,0001 is not divisible by 11.

Thus by merely looking at a number its divisibility by 11 can be determined. The student should observe the number and note "mentally" which digits are at an odd place and which at even. They should be added respectively mentally to find whether the difference between the two sums is divisible by 11 or not.

Problems :

Find divisibility or not by 11 :

P^1:   i) 341, ii) 360, iii) 5841, iv) 6234, v) 8030, vi) 8478, vii) 10,043, viii) 11,419, ix) 12,345, x) 13,394, xi) 15,565, xii) 1,001, xiii) 10,001, xiv) 1,00,001, xv) 10,00,001.

Hint: In case of numbers with a large number of zeros sandwiched between two "1"s at extreme ends as in the last 4 problems, the student should merely find out whether the last digit is at an even place or odd. If it is at an odd place, the number would never be divisible by 11 as the first digit is always at the odd place. This is simply because the sum then of oddly-placed digits would be 2 while the sum of evenly placed digits would be 0. The difference thus would not be divisible by 11 and therefore the number also would be non-divisible by 11. Similarly, if the last digit is at an even place, the number would be divisible by 11. This is because the difference between sums of digits at even and odd places would be zero. [Both these sums would be equal to 1]. "0" is divisible by 11 and so would be the number. To generalize: **If there are 'n' zeros sandwiched between '1's at the extreme ends, the number would be divisible by 11 if and only if 'n' is even; in case 'n' is odd, it would be non-divisible by 11.**

This is simply because if 'n' is even, the last digit '1' would also be at an even place. The difference between sums of digits at even and odd places would then be zero making the number divisible by 11. In case 'n' is odd, the sums of digits at even and odd places would be 0 and 2 respectively; the difference thus would be 2 which is indivisible by 11. Hence that number too would be indivisible by 11. It is to be noted in all these examples the first digit '1' would be always at the odd place.

Finally, we give two more examples illustrating this rule further. Consider the sequence of numbers:

——

$$\underset{99}{I} \quad \underset{999}{II} \quad \underset{9999}{III} \quad \underset{99999}{V} \quad \text{and} \quad \underset{999999}{VI}$$

The numbers of I, III, and VI are clearly divisible by 11. This can be readily observed by merely looking at them. Similarly, II and V are not divisible by 11. This too is evident. But why is it so?

In case of I, II and VI, the number of digits is even, being 2, 4 and 6 respectively. Hence, the sum of "9"s at even and odd places becomes equal and their difference reduces to "0". Therefore the whole number becomes divisible by 11. In case the number of "9"s is odd as in II and V, this difference becomes 9. But 9 is not divisible by 11. Hence II and V are not divisible by 11.

Hence, the rule: [**A number with even number of "9"s as digits is always divisible 11. But a number with odd number of "9" is never divisible by 11**]

Another similar example would be $\underbrace{11}_{2}$, $\underbrace{1111}_{4}$, $\underbrace{11111111}_{8}$ -- **All such numbers with even number of "1"s are divisible by 11**.

$\underbrace{111}_{3}$, $\underbrace{11111}_{5}$, $\underbrace{1111111}_{7}$, $\underbrace{111,111,111}_{9}$ -- **All such numbers with odd numbers of "1"s are not divisible by 11**.

The student should verify this by direct computation.

Here again the same rule will apply. [**Numbers with even number of "1"s as digits would always be divisible by 11. Those with odd number of "1"s as digits would never be divisible by 11**].

This is because the difference between sums of digits at even and odd places becomes "0" in case of the former and "1" in case of the latter.

What is true for "1" and "9" will also hold for a sequence of any digit between 1 to 9. We can generalize this by saying that any sequence of the same digit repeated 'n' times (like 2222, 333333, etc.) will be divisible by 11 only if 'n' is even. This is because the difference between the sums of digits at even and odd places would then be zero. In case n is odd this difference would be 2, 3..... etc., being equal to the digit that is repeated. That number therefore would be indivisible by 11.

It is to be noted this is the wonderful power of generalization in mathematics which makes it such a fascinating discipline.

Here the question will naturally arise as to the reasoning behind the above- mentioned rules relating to divisibility by 2, 3, 4... and 11. To grasp this reasoning we have to understand how our number system is built up. Take a 4-digit number with digits a, b, c, d. To illustrate: take the number 3456. Here a-3, b=4, c=5, d=6. But what is the value of this number - it is : $3\times1000+4\times100+5\times10+6$. Thus the value of the number abcd would be : $1000a+100b+10c+d$.

This is the basic principle which is the basis of our number system. It applies to all numbers, irrespective of the number of their digits. Thus a 3-digit number abc can be expressed as: $100a+10b+c$.

A five digit number abcde can be expressed as :

$10,000a+1000b+100c+10d+e$ and so on and on.

The student needs to clearly gasp this basic concept of our arithmetic. He will only then properly understand how various rules regarding testing divisibility by 2, 3, 4, 5, etc., have been formulated in Vedic mathematics. Without such an understanding, they will remain a barren set of instructions to be carried mechanically. This would block his further progress.

Let us now see how various rules regarding divisibility by 2, 3, etc., logically flow from the basic structure of our number system.

Divisibility by 2:
Let us take a five digit number abcde. It is expressed as

$$10,000a + 1000b + 100c + 10d + e$$
$$\quad I \qquad\quad II \qquad III \qquad IV \quad V$$
..... (A)

Its constituents I, II, III and IV are clearly divisibly by 2; this is because they are multiples of 10, 100, etc. Therefore, divisibility of the number depends on the divisibility of e or the last digit. Hence the rule that a number is divisible by 2 if its last digit is divisible by 2 or if it is on even number.

Divisibility by 3 :
The number abcde or 10,000a + 1000b + 100c + 10d + e can also be written as :

$$I \qquad\qquad\qquad\qquad\qquad II$$
$$[9999 \text{ or } 3333 \times 3]a + a + [333 \times 3 \text{ or } 999]b + b +$$

$$III \qquad\qquad\qquad\qquad IV \qquad\qquad V$$
$$[99 \text{ or } 33 \times 3]c + c + [9 \text{ or } 3 \times 3]d + d + e$$

The first parts of the constituents from I to IV like 9999, 999, 99 and 9 are clearly divisible by 3. It would be thus observed that the remainder by dividing the constituents I, II, III and IV is a+b+c+d. The last digit remains as it is. Therefore the remainder of the entire number after dividing by 3 : a+b+c+d+e i.e. the sum of its digits. **Hence the rule that a number is divisible by 3 only if the sum of its digits is divisible by 3.**

Divisibility by 4:
Here we go back to the break-up of a five digit number abcde which works out as:

$$I \qquad\quad II \qquad\quad III \qquad\quad IV \quad V$$
$$10,000 \times a + 1,000 \times b + 100 \times c + 10 \times d + e$$
..... (A)

It would be noted that the constituents I, II and III will be always divisible by 4 as they are multiples of 10,000 (=2500 × 4), 1,000 (=250 × 4) and 100 (=25 × 4). **Therefore the divisibility by 4 is determined only by the last two constituents or the last two digits.**

Divisibility by 5:
We again look at a five-digit number given earlier see (A). It is observed that all its constituents I, II, III and IV are always divisible by 5 being multiples of 10, 100 etc. Thus, it is only the last digit which determines whether a particular number is divisible by 5 or not. But there are only two digits which are divisible by 5 from 0 to 9 and they are 0 and 5. **Hence the rule that only those numbers which have endings as 0 or 5 are divisible by 5.**

Divisibility by 6: $6 = 2 \times 3$.
Hence a number divisible by 6 has to be divisible by both 2 and 3. Therefore only those numbers which satisfy the benchmarks applicable 2 and 3 (even numbers and sum of digits being divisible by 3) would be divisible by 6.

Divisibility by 8:
Here for clarity we take a 6 digit number (abcdef). This breaks up as :

$$\overset{I}{100,000 \times a} + \overset{II}{10,000 \times b} + \overset{III}{1000 \times c} + \overset{IV}{100 \times d} + \overset{V}{10 \times e} + \overset{VI}{f}$$
..... (B)

The first three constituents shall be always divisible by 8 being multiples of 1,00,000 (=12,500×8), 10,000 [=1250×8] and 1,000 [=125×8]. It is therefore only the last three constituents which determine whether a particular number is divisible by 8.

Hence the rule that we should look only at the last three digits to know whether a particular number is divisible by 8.

Divisibility by 9:
Again we look at the break-up of the above number abcdef at (B). It can be written as the sum of:

$$\overset{I}{99{,}999 \times a + a} + \overset{II}{9999 \times b + b} + \overset{III}{999 \times c + c} +$$

$$\overset{IV}{+ 99 \times d + d} + \overset{V}{9 \times e + e} + \overset{VI}{f}$$

It would be observed if all the constituents are divided by 9, the remainders would be a,b,c,d,e and f respectively. **Hence, the rule that if the sum of the digits is divisible by 9, the whole number is divisible by 9.**

Divisibility by 11
Here we need to do the split up of the number abcdef somewhat differently. This six digit number can also be expressed as:

$$\overset{I}{[100{,}001 \times a - a]} + \overset{II}{[9999 \times b + b]} + \overset{III}{[1001 \times c - c]} +$$

$$\overset{IV}{+[99 \times d + d]} + \overset{V}{[11 \times e - e]} + \overset{VI}{f} \qquad \text{..... (X)}$$

There is no need to be confused here: Please note:

Table A

I =	$100{,}001 \times a - a$	=	$100{,}000 \times a$
II =	$9999 \times b + b$	=	$10{,}000 \times b$
III =	$1001 \times c - c$	=	$1000 \times c$
IV =	$99 \times d + d$	=	$100 \times d$
V =	$11 \times e - e$	=	$10 \times e$

As shown on the R.H.S. (Right Hand Side) of Table A, this is how we have always given the break-up of a six-digit number like (X) as shown above. Also note:

Table B

$$
\begin{array}{lcl}
100,001 & = & 11 \times 9091 \\
9999 & = & 909 \times 11 \\
1001 & = & 91 \times 11 \\
99 & = & 9 \times 11 \\
11 & = & 1 \times 11
\end{array}
$$

The student will readily observe that all the numbers on L.H.S. (Left Hand Side) of Table B from 11, 99, 1001 to 100,001 are divisible by 11. **Therefore, if the above 6 digit number is divided by 11, the remainder would be −a+b−c+d−e+f.** How does this work out. Look at the six-digit number (X) above.

If the constituent I is divided by 11, the remainder is -a; this is because 100,001 is divisible by 11 [Table B]; similarly, if the constituent II is divided by 11, the remainder would be +b, and so on. The last digit f remains as it is. The remainder therefore becomes : [b+d+f] − [a+c+e].

If this remainder is divisible by 11, the whole number would be divisible by 11. The student needs to observe that the digits $\overset{1\ 2\ 3\ 4\ 5}{a,b\,c,d,e}$ and $\overset{6}{f}$ occur at odd and even places. It is to be noted that remainders in odd places are negative while in even places they are positive.

Hence the rule : Add the digits at even places. Add the digits at odd places. Observe the difference. If it is divisible by 11, the whole number would be divisible by 11.

Before ending this discussion of various criteria to judge whether a particular number is divisible by 2, 3, etc., we will cite the Sutras which have been applied.

Sutra: अन्त्स्यैव or Antasyaiv
This means : the last digit only.
This applies to division by 2, 5 or 10. As we saw earlier, division by 2 requires the last term to be divisible by 2 or be an even number; division by 5 calls for the last term to be 0 or 5.

Division by 10 will need the last term to be 0. This is because any multiple of 10 must have a "zero" at its end.

As regards the divisibility by 4, following is the Sutra, सोपान्त्यद्वयमन्त्यम् or Sopāntyadvayamantyam. This means : the sum of the last and twice the penultimate. Here the words mean as follows :

सोपान्त्य   =   Penultimate

अन्त्यम्   =   End

This Sutra works out as follows:

Take a number say : 3456. Its last digit is 6, twice of the second last digit becomes $5 \times 2 = 10$. The sum is 16 which is divisible by 4. Hence 3456 is divisible by 4.

But our rule was : a number is divisible by 4 only if the last two digits are divisible by 4 which in above case is $56 = 4 \times 14$.

How are these two rules to be reconciled. Take a two digit number ab. It is expressed as : $10a+b=8a+2a+b$. Division by 4 would thus leave the remainder $2a+b$. Obviously, the whole number would be divisible by 4 only if the remainder is divisible by 4 i.e. $2a+b$ is divisible by 4. But $2a+b$ is nothing but the twice the penultimate (a) and the final (b) as the Sutra lays down.

As regards 8 we saw a number is divisible by 8, if its last three digits are divisible by 8. Here the Sutra applied is : आनुरुप्येण or proportionately.

Why is this so?
We saw in case of 2, it is the last digit that matters; in case of 4, it is the last two digits; in case of 8, it is the last three digits that matter; in case of 16, it is the last four digits that have to be considered. It goes on and on in this way. Hence the Sutra uses the term proportionality.

As regards 11, the Sutra is संकलनव्यवकलनाभ्याम्

This means 'by addition and subtraction'. How is to so? This is because as we saw earlier to test the divisibility by 11, the difference (subtraction) of the sums (addition) of digits at even and odd places has to be done. Here it may be clarified that in Sanskrit :

संकलन = Addition

व्यवकलन = Subtraction

Column II is the difference between the sum of digits at even and odd places.

		Divisible by 11 or not
i)	0	✓
ii)	-3	✗
iii)	0	✓
iv)	3	✗
v)	11	✓
vi)	3	✗
vii)	0	✓
viii)	12	✗
ix)	-3	✗
x)	-4	✗
xi)	0	✓
xii)	0	✓
xiii)	2	✗
xiv)	0	✓
xv)	2	✗

Section II

Application of Vilokanam
How to find a cube root when in two digits by mere observation.

When we are asked to find a cube root of a number, in the traditional system there is only one way - it is to factorize it. For instance: 343 = 7×49 =7×7×7.

Thus, the cube root of 343 is 7. Take another number: 1728. Factorizing it:

1728 = 2×864

= 2×2×432 = 2×2×3×144

= 12×12×12.

Thus, the cube root of 1728 is 12.

But factorizing can become complex if it is a big number. Moreover, this would be far from easy if it is a cube of a prime number like 19, 37 etc. Prime numbers are those which cannot be factored. **This is because the cube of a prime would be divisible by no number except that prime. If it is a cube of 19 for example, there is no way of knowing it by the traditional method except finding out whether it is divisible by any of the previous primes i.e. 2, 3, 5, 7, 11, 13 and 17. When all this fails then only one will turn to 19. Surely, this is a laborious and time-consuming procedure.**

But the Vedic mathematics shows an easy way out if the cube root is in two digits. What is it?

We have given below a chart showing the cubes of numbers from 1 to 9. This chart goes as -

$1^3 = 1, 2^3 = 8, 3^3 = 27, 4^3 = 64, 5^3=125, 6^3=216, 7^3=343, 8^3=512, 9^3=729.$

Thus the end digit of the cube of numbers from 1 to 9 can be classified as follows:

Chart A

Number	End digit of its cube	
1	1 (1)	
4	4 (64)	
5	5 (125)	The cube is
6	6 (216)	indicated in the
7	3 (343)	bracket.
8	2 (512)	
9	9 (729)	

Thus, in the case of following five numbers the end digit is the same as the number being cubed. These are: 1, 4, 5, 6, 9

In the case of following four numbers there are alternate pairs.

Number		Last cube digit	
3	→	7 (27)	
7	→	3 (343)	The cube shown in
2	→	8 (8)	the bracket.
8	→	2 (512)	

Thus, if we are to find the cube root of a number, first we should only look at its last digit. This immediately will tell us what is the second digit of the number being cubed. It should be borne in mind that the last digits of cubes of numbers from 1 to 9 are all different. There is therefore no possibility of any error. See chart (A).

But what about the first digit of the number being cubed.

This can also be done by merely observing the number given. This is nothing but Vilokanam. We have given below cubes of numbers from 10, 20, 30 to 70, 80, 90 and 100.

Chart B

10^3	=	1,000	20^3	=	8,000
30^3	=	27,000	40^3	=	64,000
50^3	=	1,25,000	60^3	=	2,16,000
70^3	=	3,43,000	80^3	=	5,12,000
90^3	=	7,29,000	100^3	=	10,00,000

It should be noted that the values of cubes of all two digit numbers will fall between 1,000 (10^3) and 10,00,000 (100^3). Each of these cubed numbers has to belong to one of the categories as shown in Chart B. These categories are : 1,000 to 8,000, 8,000 to 27,000, 343,000 to 512,000, etc.

On the basis of Charts (A) and (B), we can determine the two-digit cube root of any number under consideration. Let us see how this works out. But here the student is advised to thoroughly familiarize himself with Charts (A) and (B).

Take the number : 39,304. What is its cube root? The end digit is 4. So the second digit of the cube root has to be 4. [See chart A]. The number is less than 64,000 but more than 27,000. Obviously, it is between 30 to 40. [See Chart B] So the first digit has to be 3.

The number is therefore 34.

Take the number 3,00,763.
What would be its cube root?
The last digit is 3. So the last digit of the cube root has to be 7 (Remember $7^3 = 343$. See chart A). As regards the first digit; the above number is between 216,000 (60^3) and 343,000 (70^3). See chart B. Thus, it is between 60 and 70. Obviously, its first digit has to be 6.

The number has to be 67.

Take one more illustration. What is the cube root of 6,36,056 ?
Last digit 6 → second digit of the cube root has to be 6 [see chart A]. This number it may be noted lies between 512000 (80^3) and

729000 (90^3). Obviously it is between 80 and 90 (see chart B). The answer is 86.

The Sutra applicable here is : लोपनस्थापनाभ्याम्
This means : By Elimination and Retention.
Here लोपन = Disappear / eliminate
स्थापनाभ्याम् = To firmly establish/retain.

In what way is this Sutra applicable to earlier discussion?
It will be recalled that just by observing the last digit, we choose (retain) one specific digit as the second digit of the cubed number and eliminate rest of the digits. Similarly, by observing the first three digits of the cube, we determine the first digit of the cubed number, ruling out others.

P^1 : Find the cube roots of following numbers:
i) 17,576, ii) 42,875, iii) 4,74,552, iv) 74,088.

This should be done by merely looking at the number. No calculation. The student should verify by direct multiplication the correctness of his answer. The student should rest assured that the author himself has solved these problems by merely observing them and without looking out at the charts (A) or (B).

Further application of the "Vilokanam" Sutra:

Divisibility by 7: As in the case of 2, 4, 5, 9 etc. there are no hard and fast rules to say whether a particular number is divisible by 7 or not. However, this can easily be inferred by merely considering some other numbers which are close to the number being considered and which are divisible by 7.

For example: Take the number 1500. Is it divisible by 7 or not? There is no need to carry out the actual division.

We need to only note that 1500 = 1400 + 100. Anybody can see that 1400 is divisible by 7 while 100 is not. Therefore, 1500 is <u>not</u> divisible by 7.

To give another example: Let us take the number 31,295. This can be split up as 28000 + 3295. But 3295 = 2800 + 495. Now, 28000 and 2800 are clearly divisible by 7 while 495 is not. Therefore, <u>31295</u> is <u>not</u> divisible by 7.

Problems:
Find out by merely looking at the following numbers whether they are divisible by 7.

P^2 : i) 7009, ii) 10,600, iii) 14367, iv) 6500, v) 7113, vi) 70,014, vii) 18335.

Note: All these problems should be solved mentally without any writing aid. The student should convince himself that his answer is correct by carrying out the division by the normal method.

The Sutra which is applied here is : संकलनव्यवकलनाभ्याम् or Sankalan Vyavakalanbhyam.

This means : By addition and substraction. When we look at proximate numbers to find whether a particular number is divisible by 7 or not, we do exactly this. For example : Take 1500. Is it divisible by 7. We substract 100 from it and get 1400. 1400 is divisible by 7 but 100 is not. So we conclude that 1500 is not divisible by 7.

This method can be applied for quick computation in several ways. To give some examples:
Suppose we have to add 531 + 692 mentally. This can be done as follows:
500 + 600 = 1100
30 + 90 = 120
2 + 1 = 3

So the answer becomes 1223.

Thus by splitting numbers process of computation becomes easy. For example: 245 - 107. Mentally this should be done as [200-100] + [45-7].

So the answer is 138.

Suppose, we have to divide 1100 by 25. First divide by 100. The answer is 11. Then multiply by 4. So the final answer of 1100/25 is 44.

Find answers to the following by mental computation only :
P³ :   i) 1009 - 203, ii) 315 + 650, iii) 1700 - 253,
         iv) 305 + 692, v) 998 + 102

P¹: i) 26 ii) 35 iii) 78 iv) 42

P²: i) No ii) No iii) No iv) No v) No vi) Yes vii) No

Nearest numbers divisible by 7 :
      i) 7,000 ii) 10,500 iii) 14,350 iv) 6,300 v) 7,000
      vi) 70,000 vii) 18,200

P³: i) 806 ii) 965 iii) 1447 iv) 997 v) 1100

Section III
 Vedic multiplication by use of a base like 100, 1000 etc.

How to carry out multiplications of two-digit, three-digit numbers mentally :
Suppose we have to find 99×91, 92×95, 105×107, 1005×1003, etc., but without any computation. By the traditional method it won't be possible. But by the Vedic methodology this proves extremely easy. For instance, answers to the above problems have been written down by the author instantaneously by merely looking at them as follows:

99×91	=	9009
92×95	=	8740
106×107	=	11,342
1005×1003	=	10,08,015

How this has been done is explained below :

When both the numbers are less than 100 :
Let us take the pair 95 and 97. First note that the product would be in 4 digits. This is simply because both the numbers are less than 100. Hence their product would be less than 10,000 (=100×100) which is of five digits. **This is the smallest 5-digit number. Hence all numbers less than 10,000 including the product of 95×97 would be in 4 digits.** Now we will see how these 4 digits are to be determined. We will divide them into two parts.

Part A: Consisting of the first two digits. ⎤ Applicable to
Part B: Consisting of the last two digits. ⎬ multiplications of
⎦ two-digit numbers
only

Here we need to introduce a term called **Complement. Complement is the difference between the base number and the numbers being multiplied.** Here our **base number is 100 as it is closest to 95 and 97.** Base numbers are always like 100, 1000, 10000 in the Vedic methodology.

Thus, the complement of 95 = 100 - 95 = 5
The complement of 97 = 100 - 97 = 3.

The rule for determining the first two digits or part A of the product is as follows:

Just subtract complement of any one the numbers being multiplied from the other one.   Rule I

Thus in the case 95×97:
Part A or first two digits :
95 – 3 = 97 – 5 = 92
So part A = 92

Part B: Here the rule is:

Multiply the complements of the two numbers. But if the product is in one digit, prefix a zero.   Rule II

Why is this so? **This is because Part B or second half of the answer has to consist of the same number of digits as the number of zeros in the base number.**

Here some elaboration is called for.

Thus, when base is 100, part B will have minimum 2 digits. When the base is 1,000, the part B will have minimum 3 digits. When it is 10,000, the part B will have minimum 4 digits. When the base is 100,000, it will have minimum 5 digits, and so on.   Rule III

To further illustrate these concepts, let us take two numbers close to 100 and multiply them. We choose 94 and 95.

94×95
: This is like $[a - b][a - c]$
$= a^2 - ab - ac + bc = a[a - b - c] + bc$

Here a = 100, b = 4, c = 5

= [100–4] × [100–5]

= 100|00 a^2

–   4|00   ab

–   5|00   ac   X

+   20   bc

The student will readily observer that the first three figures in X have two zeros in the end. **The product of the complements gets added to these two zeros in the end. This explains why part B has to consist of minimum two digits standing for two zeros when the base is 100**; Similarly, it becomes of 3 digits when the base is 1000 of 4 digits when the base is 10,000.

To give one more example:

Turning now to 95×97 discussed earlier.

Part B = 5×3 = 15.

Final answer: 95×97

= Part A and Part B
or
= 9215

To illustrate how these rules operate. See the table of figures below.

95×97
= [100–5] [100–3]
=   10,000
       –   500
       –   300
       +   15   Product of complements

This is same as :

=		9,500	First two digits = the first number or 95
−		300	First digit is the complement of the second number or 97
+		15	.....Product of the complements
		9215	

Please observe : 92 = 100 - 5 - 3
                    = 95 - 3

We thus see how by subtracting from the first number being multiplied (95) the complement of the second the first two digits of the final product are obtained. This is nothing but rule I mentioned earlier.

The last two digits or Part B is clearly seen as the product of complements. (Rule II).

We give another illustration. But here the figure reached by multiplying the complements would be in one digit and hence a "zero" will have to be prefixed.

Find 97×98

Complements:
of 97 = 3 [= Base 100 − 97]
of 98 = 2 [= Base 100 − 98]

Multiplication of Complements = 6

This is only in one digit. So prefix a zero.

Part B or last two digits of the answer : 06

Part A or first two digits : Subtract the complement of one number from the other number :
97 − 2 = 98 − 3 = 95

Hence part A = 95

Final answer : Part A and Part B :
97 × 98 = 9506

But why is it necessary to prefix a zero in part B. (Rule II).

Following chart will further clarify this :

$$\begin{array}{rl}
 & \overset{X}{97} \times \overset{Y}{98} \\
= & [100 - 3] \times [100 - 2] \\
= & 10,000
\end{array}$$

−	300	The first digit is the complement of X.
−	200	The first digit is the complement of Y.
+	6	→ Product of complements
	9700	97 This to be noted is the first number
−	200	
+	6	
	9506	The 'ten' position
		Unit position

Note : The figure obtained by the multiplication of complements was only in the unit position (6). The next position of the "tens" had to be filled. Obviously this had be zero. Hence "0" was prefixed. (Rule II).

This technique works only when both the numbers being multiplied are close to 100. With some practice the student should be able to find the answer mentally. He has to beat the calculator in this. Such is the uniqueness of the Vedic method.

Of course the same technique will work when the numbers are not proximate to 100. **The beauty of mathematics lies in the fact that its rules are applicable in general and do not depend on any particular set of numbers. The student should learn to appreciate this unique facet of mathematics. He will then only enjoy its study.**

Let us apply this technique to numbers which are not near the base.

Find $\overset{I}{65} \times \overset{II}{75}$

Complement of 65 = 35
Complement of 75 = 25
Multiplication of Complements : 875

                    10010Unit
Part B =   ↑   ↑   ↑
              8   7   5

Part A : 65 − 25 = 75 − 35 = 40

The two digits of part A are in the position of 1,000 and hundred respectively:

1000100
  ↑   ↑
  4   0

Thus, 8 of part B has to be carried over to 40 of part A. Note that "zero" and "8" need to be added as both are in the hundreds position. Part A then becomes 48.

Find answer : 4875.

Find answers to the following (use of writing material allowed) as explained above:
P^1 :   i) 75×85, ii) 72×82, iii) 63×72, iv) 83×82, v) 77×69,
            vi) 86×73

Find answers mentally :
P^2 :   i) 91 × 93, ii) 95 × 97, iii) 92 × 98, iv) 94 × 95,
            v) 96 × 98, vi) 99 × 99, vii) 82 × 95, viii) 60 × 95

We now turn to the multiplication of a pair when both the numbers are above 100. The technique is similar to that which we followed when both the numbers were less than 100. But there is a slight difference.

To illustrate. Take the pair $\overset{I}{103} \times \overset{II}{107}$

The problem is : what is I × II ?

First note that the answer would consist of 5 digits as both the numbers exceed 100. Note 100×100 = 10,000. Any number above 10,000 will have to be in 5 digits. As before we divide these 5 digits into two parts A and B.

Part A will consist of first three digits while part B will be the last two digits. How to determine part A ?

Rule IV: Simply add (and not subtract as earlier when numbers were less than 100) any one number being multiplied with the complement of the second

This can be illustrated in terms of algebra as follows:

[x+a][x+b]

$= x^2 + ax + bx + ab$

$= x\,[x+a+b] + ab \qquad\qquad L$

Here x = 100, a = 3, b = 7.
The bracketed portion above (L) is nothing but rule IV.

Our numbers are
$\overset{I}{103}$ and $\overset{II}{107}$

Complement of I = 3

Complement of II = 7

Part A or first three digits of the answer = 103+7 = 107+3 = 110 (Rule IV).
Finding last two digits : Here the same rule as earlier would apply.

Part B is the figure obtained by multiplying the complements with the proviso that a "zero" should be prefixed if the product is in a single digit. This is provided the base is 100. This is nothing but rule II enunciated earlier.

For 103×107
Part B = 3×7 = 21
Final answer = 11,021

We elucidate this by giving a break-up of the process.

103×107
= [100+3]×[100+7]
+   10,000
+   300
+   700
+   21
                 ─────────
                 10,300   → First three digits = No. I
+   700   → First digit = Complement of II
+   21   → Multiplication of complements
                 ─────────
                 11,021

We take another instance when a "zero" will have to be prefixed to part B.

Find out what is : $\overset{I}{102} \times \overset{II}{103}$

Complement of I = 2
Complement of II = 3

Product of the complements = 6

But this is in one digit. Therefore "zero" will have to be affixed to it. [Rule II]

Part B = 06

Part A : This is worked out as:
102+3 = 103+2 or 105
Find answer = 10,506.

Following table will further illustrate the process:
102×103
= [100+2]×[100+3]

```
=          10,000
+             200
+             300
+               6
          ________
          10,200   → Note: First three digits constitute No. I
+             300   → First digit = Complement of II
+               6   → Product of complements
          ________
          105/06
```

The chart above clearly brings out why a "zero" has to be prefixed to 6. [Rule II]

This method can be applied for very rapid mental calculations when both the numbers are close to 100. This is similar to the restriction when both the numbers were less than 100.

P^3: Find the answers. Calculate mentally.
i) 109×109, ii) 107×101, iii) 103×106, iv) 102×112, v) 105×117, vi) 109×111

The same method can be readily used to find products of numbers close to 1,000. This can be extended to 10,000, 100,000 etc. But we won't go into that.

But it needs to be emphasized here that methods and techniques in mathematics are applicable in general. So what holds for 100 and 1,000 as base has to hold for 1,000 and

10,000 as base. Only due to constraints of space we won't go here into higher bases like 10,000, 100,000, etc., here.

We now turn to numbers with 1,000 as base. First we take a pair with both the numbers being less than and close to 1,000.

Let us start with the pair: 997×999

Its product is 9,96,003. How is this calculation to be carried out instantaneously without any writing aid?

First, it is to be noted that all such products would be of six digits. Why? This is because they would be less than 1,000,000 (=1,000×1,000) which is the smallest 7 digit number.

As before we will get the answer in two parts, part A and part B. Part A will consist of first 3 digits and part B of the last three.

But how do we know what part A and part B are with respect to

$$\overset{\text{I}}{997} \overset{\text{II}}{\underset{\times}{}} \overset{\text{II}}{999}$$

To find part A : Here the rule is the same as applicable to numbers which had 100 as base. [Rule I]. This states: **Subtract complement of one number being multiplied from the other**

Let us see what result we get for the above pair.

Complement of 997 = 3
Complement of 999 = 1

Part A = 997 − 1 = 999 − 3 = 996.

How about part B. As earlier, it is the product of complements. In this case it becomes 3×1=3. But that is only of one digit. As a mentioned earlier part B should consist of 3 digits. This is because the base 1,000 has 3 zeros. [Rule III] Obviously two zeros will have to be prefixed to it. Thus part B is 003. Finally, 997×999 = part A and part B would be : 9,96,006.

This will become further clear from the following chart.

I × II

997×999

= [1000–3]×[1000–1]

```
=       10,00,000
–            3,000   . . . First digit complement of I
–            1,000   . . . First digit complement of II
+                3   . . . Product of complements
         ─────────
          9,97,000   → The first 3 digits = Number I
–            1,000   → The first digit = Complement of II
         ─────────
          9,96,000
+                3   → Product of complements
         ─────────
          9,96,003
```

We thus clearly see why two zeros have to be prefixed to 3.

To take another example:
To find:
995×983

We proceed as before:
Complement of 995 = 5
Complement of 983 = 17

Part A = 995–17 = 983–5 = 978
Part B = Product of Complements = 17×5 = 85.

But we need 3 digits in part B. [Rule III]
So prefix a "zero". [Rule II]
Hence part B = 085.

Final answer: Part A and Part B
= 9,78,085
= 995×983

Once the methodology is duly grasped the student should be able
to do such exercises in the split of a second.

Third example:

Find : $\overset{I}{973} \times \overset{II}{990}$

Complement of I : 27

Complement of II = 10

Part A = 973 – 10 = 990 – 27 = 963

Part B = Product of complements = 27×10 = 270

Answer : Part A and Part B

9,63,270.

P⁴: Find answers to the following -
971×998, 983×980, 960×988,
975×975, 979×991, 950×991.

All these problems to be solved mentally. Only answers to be written down.

Finally, we turn to the case when both the numbers are above 1,000. Note here that the product would be in 7 digits. This is because it would be greater than 1000×1000=1,000,000 which is in 7 digits.

Here, the part A would turn out in 4 digits and part B in 3 digits.

While calculating part A the complement of one member of the pair has to be added to the other. This is exactly similar to what we were doing when both the numbers exceeded 100.

$$\ldots\ldots \text{Rule IV}$$

To illustrate

Find : $\overset{I}{1013} \times \overset{II}{1005}$

Part A : Any of the two numbers + complement of the second :
In this case this becomes : 1013 + 5 = 1005 + 13 = 1018

This can be understood in terms of the same algebraic formulation which was given earlier.

[x+a][x+b]

= x[x+a+b] + ab   . . . . . Rule IV

Turning to part B :
Product of complements = 65

But part B has to have 3 digits when the base is 1,000. Hence one zero needs to be affixed.   . . . . . Rule III

So part B = 065.

Final answer : I × II
= 1013 × 1005
= 10,18,065

P^5 :   Solve the following : (No use of paper allowed)
i) 1010 × 1030, ii) 1053 × 1003, iii) 1007 × 1025, iv) 1050 × 1011

Finally, we turn to examples : one with base 10,000 and another with base 100,000.

Find [10,050] × [10,025]   . . . . . (i)

[Hint : Note here past B should have minimum 4 digits because the base 10,000 has 4 zeros]

Find : [100,001] × [100,001]   . . . . . (ii)

[Note : here the base is 100,000 which has 5 zeros. Hence part B should have 5 digits].

The student should get following answers.
For (i) :   10,07,51,250

For (ii) :   10,00,02,00,001

To carry out both these multiplications by the traditional method
is very time consuming with a possibility of error. **In the Vedic
system the answers can be written down straight way
without any possibility of error. Such is the glory of Vedic
mathematics.**

Finally, in passing we may mention another special feature of the
Vedic system of computation.

It enables us to conclude that a particular answer has to be
incorrect by merely glancing at the numbers concerned without
any calculation whatsoever. To give two instances:

P⁶ :   i)   65×75 = 3975 ?   Are they correct ?
       ii)   78×62 = 3936 ?

By merely looking at these numbers one should be able to say
the above answers are wrong. How is this to be inferred? This
was dealt with in an earlier part of the present section.

P^1 : i) 6375 ii) 5904 iii) 4536 iv) 6806 v) 5313 vi) 6278

P^2 : i) 8463 ii) 9215 iii) 9016 iv) 8930 v) 9408 vi) 9801 vii) 7790
viii) 5700

P^3 : i) 11,881 ii) 10,807 iii) 10,918 iv) 11,424 v) 11,235
vi) 12,099

P^4 : i) 9,69,058 ii) 9,63,340 iii) 9,48,480 iv) 9,50,625 v) 9,70,189
vi) 9,41,450

P^5 : i) 10,40,300 ii) 10,56,159 iii) 10,32,175 iv) 10,61,550

P^6 : i) and ii) are incorrect
i) 65 - 25 = 40 (25 is the complement of 75).
ii) 78 - 38 = 40 (38 is the complement of 62)

Therefore the first two digits of answers in any case cannot be less than 40. The student should recall the earlier discussion. But in the so-called answers they are shown as 39 (<40). Hence in a flash we can conclude they are wrong. The correct answers are:
i) 4875 ii) 4836

It is not enough to get the answers right mentally. The student should be able write them straight away faster than somebody with a calculator can jot them down. Then alone he can claim to have mastered the technique.

Section IV
Squaring a number and choosing a working base

We have discussed so far how numbers which are close to base numbers like 100, 1000, 10,000, etc. can be multiplied mentally without any calculation.

There is a special Sutra which is applicable only to squares of numbers. This goes as :

यावदूनं तावदूनीकृत्य वर्गं च योजयेत् । ..... (A)

तावदूनीकृत्य = Double   ऊनं = Deficiency
वर्गं = Square   योजयेत् = Attach

The meaning of the Sutra is : Whatever be the deficiency, double it; Then suffix the square of the deficiency.

To apply this in practice: Let us square 91. But 91 = 100-9.
Here deficiency is = 9.

As per the Sutra we have to further reduce the number to be squared by the some extent i.e. 9. The first part of the answer thus becomes: 91 - 9 = 82. This has to be suffixed to the first part as laid down in the Sutra under reference.

The second part is the square of this deficiency or 81.
Hence $91^2 = 8281$

In terms of algebra this Sutra is easily explained as follows-
$(x-a)^2 \quad = x^2 - 2ax + a^2$
$\quad\quad\quad\quad = x [x-2a] + a^2$

Here in the above example x = 100; a = 9.
Therefore x-2a = 100-18 = 100-9-9 = 91-9 = 82. 2a thus stands for twice the deficiency as stipulated in the Sutra. As regards multiplying by 100, this happens automatically as the first two digits i.e. 82 are placed in first and second place which is a

1000s and 100s. The same Sutra will apply even if both the numbers are above the base.

Let us square 103.
Here deficiency (excess) = 3.
Twice of it = 6.
Therefore first part of the answer = 106.

Second part = Square of the deficiency (excess) = 3×3 = 9. But we have to prefix a "0" as the second part must consist of two digits. Hence 103^2 = 10,609. **But why have we prefixed "zero"?**

The student will recall that in the earlier section (III) this rule about prefixing "0" was discussed. **The student should bear in mind that the second part of the answer must have as many digits as the number of zeros in the base number which is 100 (two zeros) here.** To clarify this further let us break up 103 × 103.

$$103 \times 103 \quad = \quad [100 + 3]\ [100+3]$$
$$= \quad 100[100+3+3]+9$$
$$= \quad 100[106]+9$$
$$= \quad 10600+9$$
Hence first part $\quad = \quad 106$
(Two zeros in the end are not shown).
Second part $\qquad = \quad 3×3=9$
But a "zero" has to be prefixed to 9.
Hence final answer : 10,609.

The algebraic equivalent of the above is : $[x+a]^2 = x^2 + 2ax + a^2$
$$= x\ [x+2a] + a^2$$

The Sutra at (A) in the beginning is very simple and elegant. It can be extended to larger numbers in the same way. For example

$989^2 \quad = \quad 988121$
$988^2 \quad = \quad 976144$
$1007^2 \quad = \quad 1014049$

$1011^2 \quad = \quad 1022121$
$115^2 \quad = \quad 13225$

The computation of 115^2 will need some elaboration.
The first part of 115^2 comes as $= 130$.
But the second part equals 225. **But the second part should have only 2 digits as the base is 100.** The student would readily recall this point has been explained repeatedly in section III.

Hence "2" on the extreme left has to be added with 0 in the first part. This can be better understood as :

First part $= 130$
But in reality it is $100 \times [100+30]$ or $130,00$
Hence final answer :

$$\begin{array}{r} 13000 \\ + \quad 225 \\ \hline 13,225 \end{array}$$

Thus, $115^2 = 13,225$

As soon as you see the number to be squared, you must be able to write its answer in a natural manner without any hesitation.

This is how we have to beat the calculator in our computation. Thus,

$10007^2 \quad = \quad 10014, 0049 \qquad$ [The second part must consist
$10015^2 \quad = \quad 10030, 0225 \qquad$ of 4 digits as the base 10,000
                                             has 4 zeros]

$100008^2 \quad = \quad 100016, 00064 \qquad$ [The base is 1,00,000 with five
                                                  zeros]

Q^1: Solve the following mentally. Answers to be written rapidly.
i) 95^2, ii) 109^2, iii) 112^2, iv) 1010^2, v) 990^2, vi) 10009^2,
vii) 9999^2, viii) 100011^2

Before we end this section we will illustrate how many multiplications can be done rapidly by the Vedic method even if the numbers are not close to 100, 1000, etc. For example, suppose we have to multiply 505 by 510. Obviously, neither 1000, nor 100 is suitable as a base. **We therefore adopt 500 as a working base.** The calculation goes as :

$$510 \times 505 = [500+10][500+5]$$
$$= 500[500+10+5]+50$$
$$= 500[515]+50 \qquad\qquad ..... (A)$$

This is precisely the same as what we were doing repeatedly earlier.

In terms of algebra:
$$[x+a][x+b] = x[x+a+b]+ab$$

Rewriting A :
$$510 \times 505 = 500[515] + 50$$

Here we have to make a slight modification in our methodology. We take 100 as our base. Therefore we write:

$$500[515] + 50$$
$$= 5 \times 100[515] + 50 \qquad\qquad ..... (B)$$

We get the interim figure as:
515/50

This is exactly what we would have got if 100 were the base. But it is not so. Therefore the first part needs to be multiplied by 5. This will be clearly seen from (B) above.

So the first part becomes:
$$2575 = 515 \times 5$$

The second part = 50
Final answer therefore = 2,57,550 = 257500+50.

We could have also taken 1000 as the base. In that case the calculation would go as follows:

500[515]+50

$$= \frac{1000}{2}\ [515]+50$$

The first part becomes: $257\frac{1}{2}$
The second part is : 10×5 = 50.
But we have to affix a zero as the second part must have 3 digits as our base of 1,000 has 3 zeros.
Hence the interim answer is:
= 257 1/2 / 050

The student should not get bewildered by 1/2 in the first part. We will need to multiply it by 1,000 (the base number) :-

$$
\begin{array}{rr}
 & 257\ 1/2 \times 1000 \\
= & 2,57,000 \\
+ & 500 \\
\hline
 & 2,57,500 \\
\end{array}
$$

But 050 is the second part.
So the final answer becomes:
2,57,500 + 050 = 2,57,550. [The second part is to be compounded with the first part as per the Sutra]

Here the Sutra being followed is: अनुरुपेण or proportionately.
In this way :
610 × 601 = 3,66,610
602 × 603 = 3,63,006

Detailed calculations are shown below :
610 × 601 = 611 / 10

Multiplying first part by 6, the final answer becomes : 3,66,610
Similarly,   602 × 603 = 605 / 6
Multiplying the first part by six, the final answer becomes: 3,63,006

[Note "zero" is prefixed to second part as it must have two digits since 100 is the base.]

In case both the numbers are less than the working base, the deficit will have to be deducted. This as per the algebraic formulation : [x-a] [x-b] = x[x-a-b] + ab

The student will recall we have done this many, many times earlier in section III. To compute in this way 695 × 699
Here our base in 100 and sub base is 700.

Interim answer : 694/5

Multiplying first part by 7 and prefixing zero to second part.

Final answer : 485805 [Note zero is affixed to second part as it must have 2 digits, 100 being the base].

Q^2 : Solve mentally: Answers to be written immediately.
i) 305×307, ii) 401×403, iii) 501×502, iv) 603×605, v) 801×802

In conclusion we may finally sum up of our discussion in both this section and the earlier one as follows:

It should be observed that both in this section as well as the earlier section III, the interim answer gets divided into two parts : first and second. It needs to noted that **the second part is compounded with the first part. As a result two, three, etc.; zeros get automatically added to it.**

To illustrate: 24/25 becomes

$$\begin{array}{r} 2400 \\ 25 \\ \hline 2425 \end{array}$$

The number of digits in the second part is equal to the number of zeros in the base number 100, 1,000 etc. This is because the base number is always the multiplier.

All this works out so beautifully only because of our zero based decimal system. The Vedic mathematicians took full advantage of it to develop a new methodology for rapid mental calculations.

However, the student should always visualize what actually happens when he applies these rules. For example, for computing : 96×95 = 9120 : he should also clearly "see" in his mind the following : 96×95 = [100-4] [100-5] = 100[100-4-5]+20 = 100[91]+20 = 9100+20 which is written as: 91/20. So final answer is 9120.

Finally, we will review once more the working of the Sutra which we had given at the beginning of this section.

What is the square of 1001.
Note: base number is 1000.
Difference = 1

By the Sutra : Double the difference between the number to be squared and the base number = 1×2=2
First Part = 1002
Second Part = Join the square of the difference = $1^2 = 1$

But we need to prefix 2 zeros as the base number 1000 has 3 zeros.

So Second Part = 001

Final answer = 1002001
But while applying the above Sutra or formula following should be visualized in mind.
$1001^2 = [1000+1][1000+1]$
$= 1,000 [1,000+2]+1$ . . . . . (A)
Note 2 is nothing but the doubling of the difference which is 1.
$= 1000 [1002]+1.$
Hence, first part = 1002.

There will be 3 zeros after it (not shown) as 1002 is to be multiplied by 1,000. Therefore the second part must have 3 digits. Hence 2 zeros have to be prefixed to 1 which is the square of the difference.

So the final answer : 1,002,001.

Thus this is the genesis of the rule: **Number of digits in the second part = number of zeros in the base number. Base number here is 1000 which has 3 zeros.**

All this must pass through the student's mind like a flash as he writes the answer by merely looking at 1001^2. With practice this will become automatic. But by following the above technique the student won't be required to memorize any formula. It will come to him naturally. He will then truly relish the study of mathematics. This is the way mathematics is to be studied. Don't apply any formula mechanically. Always "see" in your mind how it has evolved. Enjoy yourself.

Answers to Section IV

Q^1 : i) 9,025 ii) 11,881 iii) 12,544 iv) 10,20,100 v) 9,80,100
  vi) 10,01,80,081 vii) 9,99,80,001 viii) 10,00,22,00,121

Q^2 : i) 93,635 ii) 1,61,603 iii) 2,51,502 iv) 3,64,815
  v) 6,42,402

Section V
Use of "Vinculum" digits

In Vedic mathematics the Vinculum digits can be used to reduce bigger digits to smaller ones and thus to simplify calculations. But what are Vinculum digits? They share some features with negative numbers, but their concept (of Vinculum digits) is altogether different. Let us now study them.

Usually, numbers which are close to and less than 10, 20, 30, 100, 200, etc. are conveniently expressed in terms of Vinculum digits. Basically we choose as the base a number ending with zero(s) nearest to the number we want to express in terms of Vinculum digits. Take 9 for example. 10 is obviously the nearest base number. So we write 9 as: $9 = 1\overline{1}$. Similarly, 8 becomes $1\overline{2}$. 19 with 20 as base becomes $2\overline{1}$. 95 with 100 as base becomes: $10\overline{5}$ 999 with 1000 as base will be: $100\overline{1}$.

To give some further examples: $28 = 3\overline{2}$, [30-2]
$99 = 10\overline{1}$ [100−1]
$3872281 = 4\overline{1}32\overline{2}1$ [40$\overline{0}$−13, 30−2]
$29 = 3\overline{1}$ [30−1]
$57 = 6\overline{3}$ [60−3]
$181 = 2\overline{2}1$ [20−2]
$496 = 50\overline{4}$ [500−4]
$37177 = 4\overline{3}2\overline{23}$ [40−3, 200−23]

The Vinculum digits are used in the same way as normal digits. But following should be kept in mind.

$\overline{1} \times \overline{1} = 1, \overline{1} \times 1 = \overline{1}$

Additions with Vinculum digits take place in the same way as normal digits. To illustrate:

$$1\overline{5} \quad (5)$$
$$+ \quad 1\overline{4} \quad (6)$$
$$= \quad 2\overline{9} \quad (11)$$

$$1\overline{3} \quad (7)$$
$$+ \quad 1\overline{8} \quad (2)$$
$$\overline{1} \quad \text{..... Note: "}\overline{1}\text{" is the carry over.}$$
$$1\overline{1}$$

But $1\overline{1} = 9 = 7+2$

It may be wondered whether $\overline{2}\,\overline{3}$ is the same as $\overline{23}$. Both are one and the same. Let us see how this works out.

$= 5\,\overline{\overline{2}}\,\overline{\overline{3}}$

$= 5\,\overline{\overline{3}}\,7$ [Note $\overline{\overline{2}}$ becomes $\overline{\overline{3}}$].

$= 4\,7\,7$ [50–3=43]

Similarly,

$5\,\overline{\overline{23}}$

$= 4\,7\,7$ [500 – 27]

Thus we conclude : $5\,\overline{\overline{2}}\,\overline{\overline{3}} = 5\,2\,\overline{\overline{3}}$

Here the rule is : **When the digit preceding a Vinculum digit is itself a Vinculum digit it needs to be raised by 1. To give another example:**

$2\,\overline{\overline{1}}\,\overline{\overline{3}} \rightarrow = 2\,\overline{\overline{2}}\,7\,[\overline{\overline{1}} \rightarrow \overline{\overline{2}}, 10–3=7]$

$= 187\,[2\overline{\overline{2}} = 20–2 = 18]$

This is same as :

$2\,\overline{\overline{13}} = 187\,(= 200–13)$

Having seen earlier how additions with Vinculum digits take place, let us turn to their multiplications.

$$99 \times 99 = 10\overline{\overline{1}} \times 10\overline{\overline{1}}$$

$$
\begin{array}{r}
= \quad \overline{\overline{1}}01 \quad [\overline{\overline{1}} \times \overline{\overline{1}} = 1, \; 1 \times \overline{\overline{1}} = \overline{\overline{1}}\,] \\
000X \\
10\overline{\overline{1}}XX \\
\hline
10\overline{\overline{2}}01 \\
= \quad 9801
\end{array}
$$

..... (A)

It would be observed that the multiplication of (A) was much simpler than multiplying directly 99×99.
To give one more instance:

$$98 \times 97 \quad = \quad 10\overline{\overline{2}} \times 10\overline{\overline{3}}$$

$$
\begin{array}{r}
= \quad \overline{\overline{3}}06 \quad [\overline{\overline{2}} \times \overline{\overline{3}} = 6, \; 1 \times \overline{\overline{3}} = \overline{\overline{3}}\,] \\
000x \\
10\overline{2}xx \\
\hline
10\overline{5}06 \\
= \quad 9506
\end{array}
$$

The advantage of Vinculum digits will become clearer in case of larger numbers. For example:

$$
\begin{array}{r}
299 \\
\times \quad 387 \quad \ldots\ldots (A) \\
\hline
\end{array}
$$

$$
\begin{array}{r}
= \quad 30\overline{1} \\
\times \quad 4\overline{13} \\
\hline
\end{array}
$$

(Note: $4\overline{13} = 4\overline{1}\ \overline{3}$. Hence we multiply by $\overline{3}$ and $\overline{1}$ separately.)

Thus, to find (A):

$$
\begin{array}{r}
3\ 0\ \overline{1} \\
\times \quad 4\ \overline{1}\ \overline{3} \\
\hline
= \quad \overline{9}\ 0\ \overline{3} \\
\overline{3}\ 0\ 1\ x \\
\hline
\end{array}
$$

$$\frac{\begin{array}{c} 1\,2\,0\,\overline{4}\,x\,x \\ \overline{1} \end{array}}{1\,2\,\overline{4}\,\overline{3}1\,3}$$

$$= \quad 12\overline{43}13$$

$$= \quad 1,15,713 \qquad [12\overline{43} = 1200 - 43 = 1157]$$

$$= \quad 299 \times 387 \qquad \text{(The student should check this by direct multiplication)}$$

In succeeding sections we will see further application of these digits.

However, before we end this section it would be relevant to point out the basic difference between imaginary numbers and Vinculum digits.

The imaginary numbers like -1, -2, etc. exist on their own and are on the same par as natural numbers like 1, 2, 3 etc. But Vinculum digits have no separate existence.

For example: $\overline{\overline{1}}, \overline{\overline{2}}$ etc., make no sense. **They make sense only when they are prefixed by a natural number.** Therefore,

$$1\overline{\overline{1}} = 9,\ 3\overline{\overline{2}} = 28,\ \text{etc.}$$

This is the basic and crucial difference between imaginary numbers and Vinculum digits. The latter merely represent a technical device to simplify computations; they have no role other than that. But, on the other hand, the entire foundation of modern mathematics will collapse without imaginary numbers.

P^1 : Solve using Vinculums:
i) 39×39 ii) 28×29 iii) 69×69 iv) 79×79
P^2 : Solve mentally using Vinculums:
i) 95^2 ii) 97^2 iii) 94^2

P^1:

i) $\quad 39 \times 39 = 4\overline{1} \times 4\overline{1} = 16\overline{8}1 = 1521$

ii) $\quad 28 \times 29 = 3\overline{2} \times 3\overline{1} = 9\overline{9}2 = 812$

iii) $\quad 69 \times 69 = 7\overline{1} \times 7\overline{1} = 48\overline{4}1 = 4761 \ [49, \overline{14} \rightarrow 48\overline{4}]$

iv) $\quad 79 \times 79 = 8\overline{1} \times 8\overline{1} = 63\overline{6}1 = 6241 \ [64\ \overline{16} \rightarrow 63\overline{6}]$

IMP: Problems (iii) and (iv) are to be specially noted. During multiplication vinculums are to be treated as normal numbers except that they are negative and the carryover of $\overline{1}$ is compounded (subtracted) from the digit on the left.

P^2:

i) $\quad 95^2 = 1\overline{10}25 = 9025$

ii) $\quad 97^2 = 10\overline{6}09 = 9409$

iii) $\quad 94^2 = 1\overline{1}236 = 8836$

Hint : Use the method explained chapter IV. As is observed above, it applies equally well to Vinculums as well as normal numbers.

Section VI
Vedic Subtraction without a 'carry over'

Zero (0) is perhaps the most wonderful digit in our number system. It is the basis of our 10-based numbers. By mere addition of a zero, numbers increase rapidly from 10 to 100 to 1,000 and so on.

The numbers 10, 100, 1,000, 10,000 etc. are called base numbers. They form the bedrock of the Vedic mathematics so to say. The Vedic mathematicians played with the versatility of zero to turn an ordinary subtraction into an elegant art. Let us see how they did it.

We start with the Sutra:
निखिलं नवतश्चरमं दशतः
निखिलं = All, चरमं = End, नवतः = Nine, दशतः = Ten

This means: "All digits to be deducted from 9, the last one alone from 10."
A word here would be perhaps in order for those who are interested in Sanskrit. Both the words नवतः and दशतः in the aforesaid stanza are in पंचमी conjugation which gives the sense of "from". **It needs to be emphasized here this Sutra is applicable only for subtractions from base numbers.**

Let us now put this Sutra into practice. Suppose we want to carry out the following computation.

$$
\begin{array}{rl}
954 & \ldots\ldots \text{ i} \\
- \quad 367 & \ldots\ldots \text{ ii} \qquad \ldots\ldots \text{ (A)} \\
\hline
\end{array}
$$

It works out as by the normal method as follows:

$$
\begin{array}{rl}
954 & \\
- \quad 367 & \\
\underline{11} & \\
587 & \ldots\ldots \text{ (B)} \\
\end{array}
$$

In the above Sutra there is no requirement of "carry forward 1"
as we did earlier. Instead we proceed as follows:
Take a base number nearest to 954. Obviously it is 1,000.
Subtract 367 from the base number. **To do this we apply the
Sutra : subtract the last digit from 10 and the rest from 9.**
Thus,

$$
\begin{array}{ll}
1000 & \quad 10 - 7 \;=\; 3 \\
-\underline{367} & \quad 9 - 6 \;=\; 3 \\
633 & \quad 9 - 3 \;=\; 6
\end{array}
$$

Now we add 633 to the number at (i) in the aforesaid problem
which is 954.
This works out as:

$$
\begin{array}{l}
954 \\
+\;\underline{633} \\
1587 \qquad ..... (\bar{L})
\end{array}
$$

We have to delete the first digit "1" in $(\bar{L})$. The three remaining
digits are our answer:
So 954 − 367 = 587.
The student will observe this is same as the answer at (B),
worked out by the traditional method.

Surely this was simpler and more interesting than the normal
method of subtraction involving "carry over".

Take another example:

$$
\begin{array}{ll}
1323 & \quad ..... \text{i} \\
-\;\underline{968} & \quad ..... \text{ii} \\
355 & \quad ..... \text{iii}
\end{array}
$$

Let us see how Vedic mathematics yields the same answer but
more elegantly.

First we choose a base number for 'i'. **It must have as many
zeros as the digits in the number from which subtraction is to
be carried out.** This number at 'i' is 1323 having 4 digits. Hence
our base number becomes 10,000 with four zeros.

The number to be subtracted i.e. 968 has 3 digits. To apply the Sutra we have to turn it into 4 digits. This is because our base number (10,000) has 4 zeros. This is done as : 0968 :

$$\begin{array}{r} 10,000 \\ -\ \underline{\quad 0968} \\ 9,032 \end{array}$$

This has been computed by the Sutra as follows:

$$2 = 10{-}8$$
$$3 = 9{-}6$$
$$0 = 9{-}9$$
$$9 = 9{-}0$$

As mentioned in the first example we have to add 9032 to 1323 (the number at i) and then delete the first digit 1 from the total. Let us see how this works out.

$$\begin{array}{r} 1,323 \\ +\ \underline{\quad 9,032} \\ 10,355 \quad \dots\dots \bar{\bar{L}} \end{array}$$

Deleting the first digit 1, from $\bar{\bar{L}}$ it becomes:
0355 which is the same as : 355. This was precisely earlier our answer (see iii).

Addition is simpler than subtraction. If both the numbers are 8/9 digits long, the calculation becomes much simpler by the Vedic method. To illustrate: Suppose we have to subtract 98,76,578 from 1,11,25,671.

Let us see how it works out by the normal way:

$$\begin{array}{rl} 1,11,25,671 & \dots.. \text{ i} \\ -\ \underline{\quad 98,76,578} & \dots.. \text{ ii} \\ 12,49,093 & \dots.. \text{ iii} \end{array}$$

In such a calculation there is every possibility of a mistake.

Let us see the Vedic way:
Base number = 10,00,00,000   IV

As noted earlier, the base number must have the same number of zeros as the number of digits in the number from which subtraction is made. i has 8 digits and hence we choose 10,00,00,000 (8 zeros) as the base.

Subtracting ii from the base by applying the Sutra:

$$
\begin{array}{r}
10,00,00,000 \\
-\quad 0,98,76,578 \\
\hline
9,01,23,422 \quad \text{..... IV}
\end{array}
$$

From right to left:
9 = 9–0, 0=9–9, 1=9–8, 2=9–7, 3=9–6, 4=9–5, 2=9–7, 2=10–8

Note: Unlike the traditional way, the subtraction can be carried out rapidly on the basis of the Sutra without least confusion or possibility of an error.

Now, we add i with iv.
Thus,

$$
\begin{array}{r}
11125671 \\
+\quad 90123422 \\
\hline
10,12,49,093 \quad \overset{=}{\underset{L}{}}
\end{array}
$$

Deleting the first digit 1 of $\overset{=}{L}$ our answer becomes:

$$
\begin{array}{r}
01249093 \\
\text{or} \quad 12,49,093 \quad \text{...V}
\end{array}
$$

Observe iii = V

Surely, it will be observed subtraction from the base number by the Sutra is much easier than a normal subtraction as the former involves no carry forwards. The possibility of error is

therefore substantially reduced. This is to the credit of Vedic mathematicians.

The Vedic method of subtraction certainly appears different from the normal way of subtraction. The student will no doubt wonder as to how it works. This is explained below.

The basic problem in terms of algebra can be stated as : To find a – b.

This can also be written as
$[\,(a+c)-b\,]-c$ ...Z
$=[\,(c-b)+a\,]-c$
$=[\,c+(a-b)\,]-c$

Now here c becomes the base number.

What we do in the Vedic methodology is as follows:
We first carry out c-b. We then add "a" to it. Thus, a+(c-b) or c+(a-b) becomes our interim answer ..... Y

We get the desired answer by deleting "1" which always comes in the beginning in the interim answer at Y. How this happens is explained subsequently.

We illustrate this with reference to our first problem : To calculate : $\overset{a}{954} - \overset{b}{367}$

Here c is the base number or 1,000.

So [954 - 367] = [(954 + 1000) - 367] - 1,000.

$$=\left[\left(\overset{I}{1000-367}\right)+\overset{II}{954}\right]-1{,}000 \qquad .....(A)$$
$=[\,(c\quad-\quad b)\quad+\quad a\,]-c$

Now I = 1000 - 367. This we calculate by applying the Sutra.

The answer comes = 633 [9 - 3, 9 - 6, 10 - 7]

So A becomes :

$$[633 + 954] \overset{\bar{X}}{} - 1,000 \qquad \qquad \dots \dots \bar{A}$$

It will be recalled we precisely carried out additions as in $\bar{X}$ above earlier. See $\bar{L}$, $\bar{\bar{L}}$ and $\bar{\bar{\bar{L}}}$.

The student may point out that we did not carry out deduction of 1,000 as stated in $\bar{A}$. But we very much did so.

This works out as follows. In every case we deleted "1". See $\bar{L}$, $\bar{\bar{L}}$ and $\bar{\bar{\bar{L}}}$. But, what was "1"? It stood for 1,000 or 10,000 and 1000,00,000 respectively which were our base numbers. To illustrate : we go to our first problem. Our interim answer was : 1587. We deleted "1". But what did we do in reality ? We deducted 1,000 which was our base number. This is explained below:

1587-1000 = 587 (The first digit "1" is deleted).

Finally, as we have already noted in the beginning, the Sutra quoted there was a technique for subtraction which was applicable only to base numbers. But what about the "carry forwards" of the traditional system - the student will ask.

They are very much there, though in a different form. To illustrate :

$$
\begin{array}{r}
1,000 \\
-\ 363 \\
11 \\
\hline
637
\end{array}
$$

This is the normal way. Carry forwards are also shown in the computation.

By the Sutra:

$$7 \ = \ 10 - 7$$
$$3 \ = \ 9 - 6 \ [= (10\!-\!1) - 6]$$
$$6 \ = \ 9 - 3 \ [= (10\!-\!1) - 3]$$

All the digits (except the last) are deducted from 9 in the Vedic method. But what is '9' ? It is nothing but 10-1. Thus '1' is the 'carry forward' of the traditional system. As regards the last digit, in both the systems it is deducted from 10. **Thus, the carry forward '1' very much deducted though in a disguised way in the Vedic method.**

Problem: Finally we pose a problem. It would be observed in all our interim answers $\dfrac{\overline{}}{L}, \dfrac{=}{L}, \dfrac{\equiv}{L}$ **the first digit was invariably "1". This will always happen. Why is it so?**

We have to explain why in our methodology the interim answer invariably begins with 1. First note our base numbers always begin with 1, be they 100 1,000 or 1,00,000. The base number chosen always has one digit more than the number from which subtraction is made. This is because the base number has as many zeros as the digits in the number to be subtracted from. For instance, in our first example this binary was : 1,000 and 954 in the second 10,000 and 1323. The additional digit is 'one' in the beginning.

Turn now to equations Z and Y on an earlier page. Using the same notation our desired answer is a-b.

Our interim answer: c+(a-b)or Y.

Now "a" has to have one digit less than c (the base number). Therefore (a-b) must have at least one digit less than c.

Take any base number and add any number however big having one digit less. The first digit "1" of the base number won't be affected. For example: 100+99 = 199 (1 unchanged), 100,000+99,999=1,99,999 (1 unchanged).

This is precisely what happens in our methodology. This is why our interim answers always begin with "1". (See $\dfrac{-}{L}, \dfrac{=}{L}, \dfrac{\equiv}{L}$) Thus by merely deleting "1" at the beginning in the interim answer, the base number gets deleted. This is because deletion of "1" constitutes subtraction of the base number automatically.

The student may not readily grasp the contents of the above explanation. But he should read it over carefully again and again; and suddenly the author is sure the truth will dawn upon him to his great elation like the sun emerging from clouds on a dark gloomy evening.

Section VII
How to determine prime numbers from 1 to 100.

We know that prime numbers are those which are not divisible by any other number. Thus 2, 3, 5, 7, 11, 13 etc. are prime numbers.

To identify prime numbers from 1 to 100 we have to eliminate those which are divisible by other numbers. To do this :

We first write numbers from 1 to 100 in a chart as follows:

(I)

1,	2,	3,	4,	5,	6,	7,	8,	9,	10
11,	12,	13,	14,	15,	16,	17,	18,	19,	20
21,	22,	23,	24,	25,	26,	27,	28,	29,	30
31,	32,	33,	34,	35,	36,	37,	38,	39,	40
41,	42,	43,	44,	45,	46,	47,	48,	49,	50
51,	52,	53,	54,	55,	56,	57,	58,	59,	60
61,	62,	63,	64,	65,	66,	67,	68,	69,	70
71,	72,	73,	74,	75,	76,	77,	78,	79,	80
81,	82,	83,	84,	85,	86,	87,	88,	89,	90
91,	92,	93,	94,	95,	96,	97,	98,	99,	100

We will now eliminate those which are divisible by 2, 3, 5 and 7. The student can do this easily by following rules given in earlier chapters to determine which numbers are divisible by 2, 3, 5, etc., and which not. Here again merely looking at the numbers should suffice. (Vilokanam).

The chart I then becomes :
1, 2, 3, 5, 7
11, 13, 17, 19
23, 29
31, 37, 41, 43, 47
53, 59, 61, 67
71, 79
83, 89, 97

These are the prime numbers from 1 to 100. For those who are historically minded a Greek mathematician ERATOSTHANES (276 to 194 B.C.) had made such a grid to determine prime numbers from 1 to 100.

Finally, the student may ask, or rather should ask, why we have checked divisibility by only 2, 3, 5, 7 and not by 11, 13, etc., which are the succeeding primes. This is explained as follows. All the multiples of 11 from 1 to 9 get eliminated because they are divisible by 2, 3, 5 or 7 which we have covered. The smallest multiple of 11 in which 2, 3, 5 or 7 is not the a factor is 11×11=121. But it is beyond 100.

Problem:
Identify prime numbers from 100 to 200. [Hint: Here apart from 2, 3, 5 and 7 also take into account 11 and 13. We need not bother about 17 because its smallest multiple not covered by above primes is 17×17=289. This is not to be considered being more than 200.]

Primes from 100 to 200 :
101, 103, 107, 109, 113, 115, 127, 131, 137, 139, 149, 151, 157, 163, 173, 179, 181, 191, 193, 197, 199

The student may note here the following:

i) The number of primes from 1 to 100 is 25 while from 100 to 200 it is 22. There is moreover no regularity or pattern whatsoever in their occurrence. They are spread erratically in the number system.

ii) No formula has been devised so far to determine primes between any two numbers, say, between 1,000 to 2,000. Primes occupy an important position in the 'Number Theory' but they also pose intractable problems many of which are yet to be solved

Section VIII

Vedic Division

The Vedic mathematics provides us with a very elegant methodology to carry out division. This is especially recommended when we are dividing by 9. By this method when the divisor is 9, the quotient and remainder can be written down through Vilokanam (mere observation) instantaneously in many cases.

For example:

9]1111[:   Quotient (Q) = 1 2 3
    4   Remainder (R) = 4

9]111111[:   Q = 1 2 3 4 5.   R = 6
     6

9]2112[:   Q = 2 3 4   R = 6

First, we will explain how Vedic division is carried out. This method of division is totally different from the normal method. In the Vedic system the division commences from the right while in the traditional process it starts from the left. In the Vedic division the remainder pops out in the beginning while by the traditional way it emerges in the end. **It is the dividend that becomes known first when the division is carried out normally. On the other hand, in the Vedic division both the dividend and remainder are found through addition which has no parallel in the normal process of division.** Surely the student must be curious by now to know more about the Vedic methodology of division. In what follows this has been explained in depth.

Let us start with the division of 5111 by 9, as per the Vedic rules. Here the last digit becomes a part of the remainder and hence 5111 would be written as : 511/1. To divide this number by 9, we have to take steps as mentioned:

Step 1: Bring 5 down and write it below the dividend. It becomes the first digit of the quotient.

So 9]511/1[is written
as
9]511/1[
Q = 5 .....

Step II : **Add 5 to the next digit in the dividend. This becomes 5+1=6. This becomes next digit of the quotient.**

9]511/1[
Q = 56...

Step 3 : **Repeat the same process i.e. add 6 to the next digit which is 1. So the third digit of our quotient becomes 7.** Thus

9]511/1[
Q = 567/1

To get the remainder we have to add **the last digit of the quotient which is 7 to the last digit of the dividend which is 1.** Thus the remainder becomes : 7+1=8

Thus   Quotient = 5 6 7
        Remainder = 8

Thus is how the Vedic division is carried out through addition.

The Sutra which applies to Vedic division is as follows:
Paravartya Yojayet : परावर्त्य योजयेत्

The word "Paravartya" means transpose while the word "Yojayet" means rearrange.

The student will readily observe that this is what we have been doing. When we bring down the digits one by one we transpose. When we add them to the next digit of the dividend we rearrange.

It sounds no doubt quite complicated at the first sight but it is not at all so. All what you have to do to find the quotient when the divisor is 9 is to go on adding the digits of the dividend successively until you come to its second last digit. This gives us the quotient. ...... Rule I

To get the remainder we merely add all the digits of the dividend. -- Rule II

We will give two more examples to illustrate how by applying rule I division of even big numbers by 9 can be carried out, and that too mentally with utmost case.

9] 2 3 1 5
By adding the successive digits until the second last one i.e. 1, we get the figure : 256.

But the remainder which is the sum of all the digits of the dividend is equal to 11. This is more than 9. So dividing it by 9 we get : $9 \times 1 + 2$.

To the figure 256 obtained earlier we have to add one. Hence
Q = 256 + 1 = 257
R = 2

Another interesting example would be : To divide by 9
111, 111, 111 [Taken 9 times].
To do this by traditional method would be most cumbersome and in all probability could lead to errors. But here the Vedic method yields the answer instantaneously :

The interim quotient becomes : 12345678 (merely adding the successive digits until the 8th).

The remainder is
= 9 [Addition of all digits]
But 9 = $9 \times 1 + 0$
So the final answer:
Quotient = 12345678+1

= 12345679

R = 0

Once the basic principle of the Vedic division by 9 is grasped, the student will be able carry it out the division by 9 of 111 111 111 mentally. He can straight away write the answer without any strain. He will enjoy doing this computation mentally, and thus appreciate how some problems can be tackled by Vedic mathematics very elegantly and with utmost ease.

We are sure most students will find this process fascinating. We would advise the student to confirm the correctness of the answer by direct multiplication in a normal way.

Problems: (Answers at the end of the chapter)
P^1: Divide by 9. Solve mentally.
    i) 9]8,000, ii) 9]7,000, iii) 9]600,000

(Note the symmetry and repetition of the same digit in the quotient and remainder of above numbers)

P^2:   Solve mentally:
    i) 9]1341, ii) 9]2311, iii) 9]711, iv) 9]1511, v) 9]2131

The student would have noticed that so far we have been choosing dividends in which all the digits are generally small. Of course, the same method applies when the digits are bigger. **The student should always bear in mind that in mathematics there are no exceptions. A correct method will always yield correct results, no matter whether the digits are big or small.** We illustrate this with some examples:

$$9]3956[$$

To be written as 9]395/6
Adding the digits successively until the second-last digit:
IQ = 3/12/17   (IQ stands for interim quotient.)
IR = 23   IR is interim remainder.

The dividend is of four digits. So the quotient has to be of three digits. But our IQ is 3/12/17. How has this happened?

It needs to be noted that '12' and '17' belong to the tenth and unit position respectively. We rewrite IQ as follows : 3/12/17

```
         100  10  unit
IQ =      ↑    ↑    ↑
          3   12   17
```
It now becomes:

```
          3+1   2+1    7
or         4     3     7        [By horizontal addition].
```

Thus, our interim quotient is
= 437

Why do we still say "interim"? This is because the interim remainder is more than 9. IR = 23 = 9×2+5. The quotient is finally worked out only after the remainder is calculated.

We have to add "2" to the interim quotient which is 437.

So now the answer becomes :
Q = 437 + 2
= 439
R = 5

To take another example :
9]5193

By addition of digits until the second last one or 9, the interim quotient becomes:

IQ = 5 6 15
Interim Remainder = 18 (addition of digits)
But IQ = 5 6 15

This becomes by horizontal addition:
5 7 5

Interim Remainder = 18 = 9×2+0

Final answer :
Q = 575 + 2
= 577
R = 0

This might appear cumbersome. But once the student grasps how the "horizontal addition" is to be carried out, it goes quite smoothly. The possibility of error is eliminated even in case of big numbers. Take for instance :

$$9]78,956[$$

Interim quotient = 7 15 24 29
Interim remainder = 35
Consider now :
7 15 24 29

This becomes : 7+1, 5+2, 4+2, 9
Thus interim quotient : 8769
Interim remainder = 35
= 9×3+8

We therefore add 3 to the interim quotient. We get :
Q = 8769 + 3
= 8772
R = 8

Finally, to explain how the "horizontal addition" works.
Consider,
9]8888

The interim quotient is : 8/16/24

100 10 *unit*
 ↓   ↓   ↓
 8   16   24

But what do these 3 figures stand for ? Note "8" is in the 100th position, 16 in the 10th position and 24 in the unit position.

So   8 16 24
is in fact

                800
+   160
+   24
             ____
                984

This is nothing but :

        8   16   24
↓   8+1   6+2   4
↓   9   8   4

Here the interim remainder obtained by adding together all the digits is:

IR = 32
= 9×3+5

Final answer :
Q = 984+3
= 987
R = 5.

In the end we will solve fully one problem to illustrate the power of Vedic division when the divisor is 9. We have to find the quotient and remainder in:

9] 11<u>1 111 111 11</u>1
         12 times

By the traditional method the process is long and cumbersome with every possibility of a mistake. By the Vedic method we will get the result in 2/3 minutes without any strain or possibility of an error.

First, write the dividend as :
9] 111 111 111 11/1

The interim quotient becomes:
IQ = 1234567891011 (By adding successive digits till second last)
IR = 12 (sum of digits).

This is exactly what we have been doing so far.

But there is a problem here. The last two numbers are 10, and 11. What do we do with them.

Not to worry. We have to use here horizontal addition. For facility we, take only last 4 digits of the IQ and rearrange them.

They are
8, 9, 10, 11

By successively applying horizontal addition this set becomes:

$$8\ 9\ 11\ 1$$
$$\downarrow$$

8 10 11   In case horizontal addition
$$\downarrow$$ always start from the right
9 0 11   side.

So our IQ is = 1 2 3 4 5 6 7 9 0 11

But IR = 12 = $9 \times 1 + 3$.

We add 1 to IQ.

So our final quotient becomes : IQ + 1
= 1 2 3 4 5 6 7 9 0 12

Thus we get the answer :
Q = 1 2 3 4 5 6 7 9 0 1 2
R = 3

Once the student gets familiar with the process he can carry it out smoothly, quickly and enjoyably.

Finally, we pose the following problems as a mathematical riddle.
We know :
9] 111 111 111   = 1 2 3 4 5 6 7 9   ..... (A)
   (9 times)   R = 0
9] 111 111 111 111   = 1 2 3 4 5 6 7 9 0 1 2
   (12 times)   R = 3   ..... (B)

The question is : How do we deduce (B) from (A). The only difference between the two dividends is the addition of 111 of the end. But 9] 111 = 12 + R = 3   ..... (C)

How do we link (A), (B), and (C). Obviously by merely putting "12" at the end of (A) would not give the correct answer. What should we do ? We need to understand :

$$\overset{A}{} \qquad \overset{B}{}$$

111   111   111   111 = 111 $\underbrace{111\ 111\ 000}$ + 111
     (12 times)   (9 times)

9] $\overline{A}$ = 12345679000 ..... $\overline{C}$

9] $\overline{B}$ = 12   ..... $\overline{D}$ R = 3

$\overline{C} + \overline{D}$ = 12345679012 = 9] 111 $\underbrace{111\ 111\ 1111}$
                             12 times

R = 3
We know this is the correct answer at (B).
In the same way we will solve:
9] 111 $\underbrace{111\ 111\ 111\ 111\ 111}$ X
         18 times

We know 9] 1$\underbrace{11\ 111\ 11}$1 = 12345679   ..... Y
        9 times

You have to find quickly X from Y.

$$\underbrace{111, 111, \ldots 111}_{18 \text{ times}} \overset{L}{=} \underbrace{111\ 111\ 111}_{9 \text{ times}}\overset{M}{\ } \underbrace{000, 000, 000}_{9 \text{ times}} +$$

Note:

$$\underset{9 \text{ times}}{\underbrace{+\ 111, 111, 111}}^{\displaystyle N}$$

How do we work this out?

First observe:

9] L = 9] M + 9] N

But   9]M = 12345679$\underbrace{000000000}_{9 \text{ zeros}}$ [From A]

           9]N = 12345679   [Also from A]

Hence

           9]L = 9]N+9]N

             = 12345679012345679   C

This is our answer.

Note how the quotient above consists of two identical sets of digits. Observe the beauty of this sequence. Solve directly also by horizontal addition. It is no doubt a little tough. But if you can manage, rest assured that you have mastered the technique of horizontal addition. Confirm that the answer you get is the same as (C).

Finally, we will illustrate how the answer (C) can be reached in another way. We saw:

9] $\underbrace{111\ 111 \ldots\ldots 111}_{18 \text{ times}}$ = 12345679012345679   ..... (C)

We have shown subsequently :

99] $\underbrace{111\ 111 \ldots\ldots 111}_{18 \text{ times}}$ = 1122334455667789   ..... (D)

Can you deduce (C) from (D)? It is seen

11]C = D
or C = D×11

We will advise the student to do both the computations by conventional method. They are long-winded but they would help to build his self-confidence.

There is an easier way to compute D×11. This is illustrated below.

D×11 = D×10+D
This becomes
```
        1 1 2 2 3 3 4 4 5 5 6 6 7 7 8 9 0
    +     1 1 2 2 3 3 4 4 5 5 6 6 7 7 8 9
                  1 1 1 1 1 1 1 1
      ─────────────────────────────────────
        1 2 3 4 5 6 7 9 0 1 2 3 4 5 6 7 9
```

= (C)

Thus our answer at (C) was correct. Always remember addition is easier than multiplication as seen above.

P^3:   Solve the following, use of paper allowed:
    i) 9]9856, ii) 9]9356
    iii) 9]7539, iv) 9]69956

We will finally explain why and how the Vedic division works. It would baffle anybody how by successively adding the digits of a dividend the quotient is obtained; similarly, why does the remainder become the sum of digits of the dividend? These questions arise because the methodology of Vedic division is altogether different from the traditional way.

The answer lies in the beauty and symmetry of our decimal system. Let us take a four digit number : abcd

We are going to divide abcd by 9; a, b, c and d are its digits. As we have done many times before; abcd stands for

1000a + 100b + 10c + d   .....   (X)

This is to be divided by 9. Let us first squeeze out the remainder. In the Vedic system as we noted earlier remainder is nothing but the sum of digits when the divisor is 9. So here
R = a+b+c+d.

Subtracting it from the original number of (x), our dividend becomes:
999a + 99b + 9c   .....   (Y)
Note: 999a = 900a + 90a + 9a
99b =   90b + 9b
9 c =   9c

Thus, Y can be rewritten as :
900a + 90 [a+b] + 9 [a+b+c]   .....   (Z)

Now if we divide this by 9 we get the answer as :
100a + 10[a+b]+[a+b+c]

But how would this three-digit number turn out : this will simply become :

$$
\begin{array}{ccc}
100 & 10 & unit \\
\updownarrow & \downarrow & \downarrow \\
a & (a+b) & (a+b+c)
\end{array}
\qquad ..... \quad (K)
$$

Thus this is the quotient we get by dividing (X) by 9 having first subtracted from it the remainder a+b+c+d (sum of digits).

But what is (K), the quotient given above.

The student will readily observe it is nothing but the sum of digits carried out one by one until the second last digit ("c" in this case).

We need hardly reiterate this is nothing but our rule I.

The method of Vedic division described above can be applied when the divisor 8 or 7. But then the process becomes more complicated and cumbersome. It is not recommended for practical use. **It is only when the divisor is 9 and the dividend consists of small digits that the division can be carried out smoothly and by 'Vilokanam' (mere observation).**

Now we will explain how the Vedic process of division is carried out when the divisor is 8 or 7. We start with 8.

Here for 8 the base is 10. Hence the complement is 2. Let us divide any four digit number say 3456 by 8. We write this as
8]345/6[..... A

As before we take the first digit "3" down. It becomes the first digit of the quotient. So A becomes

8]345/6[

Q = 3 ...

Now since the complement is 2. We have to multiply 3 by 2 and add it to "4" in the dividend. Thus, our quotient now becomes
Q = 3 10

But again we repeat the process. (10×2+5) So
Q = 3 10 25
Carrying out horizontal division, we get : Interim Q = 425

The remainder becomes
25×2+6 = 56.

Dividing this by 8 we find
56 = 8×7+0

So final answer = 425+7 = 432
R = 0

The rule for 8 can be formulated as : **Add successive digits until the second last; but before adding any digit the preceding number in the quotient should be multiplied by 2. The remainder is obtained by multiplying the final figure by 2 and adding it to the last digit. ....** Rule III

We will illustrate this by giving one more example.
We divide :1234 by 8.
So   8]123/4[

By applying rule III given earlier, we get :
Q = 1(1×2+2), (4×2+3)
or 1   4   11
By horizontal addition
Interim Q = 1 5 1.
Interim R = 11×2+4 = 26
26 = 8×3+2
So the final answer :
Q = 154   R = 2

Finally, to explain how this rather complicated process works.

First we write the four digit number as : abcd.

This is nothing but : $1000a + 100b + 10c + d$ ..... (O)

The quotient as per rule III given earlier would be :

$$
\begin{array}{ccc}
100 & 10 & unit \\
\downarrow & \downarrow & \downarrow \\
a & 2a + b & 4a + 2b + c
\end{array}
$$
          ..... (X)

The remainder as per rule III
$R = 2[4a+2b+c]+d$
$= 8a+4b+2c+d$ ..... (Y)

Let us see how all this adds up. As seen earlier from (X),

$$Q = \begin{array}{ccc} 100 & 10 & unit \\ \downarrow & \downarrow & \downarrow \\ a & 2a+b & 4a+2b+c \end{array}$$

But what would this number be?
It would be :
$Q = 100a+10[2a+b]+4a+2b+c$
$= 124a+12b+c$

Thus this is our quotient. To get the original number we have to multiply it by 8.
$Q\times8 = 8[124a+12b+c]$
$= 992a+96b+8c$,,,,, (Z)

But our remainder is :
$R = 8a+4b+2c+d$ (see Y)

Thus,

$$\begin{array}{rcl} Q\times8+R & = & 992a+96b+8c \quad\text{,,,,, (Z)} \\ + & & \underline{8a+4b+2c+d} \\ & = & 1000a+100b+10c+d \end{array}$$

This is nothing but our original number of (O).

We finally turn to division by 7. Here the procedure is exactly the same as for 8. But there is one difference. It would be observed that the complement for 7 is 3 [=10-7] (as against 2 for 8).

We therefore multiply each digit of the quotient by 3 before adding it to the next one. This continues until the second last digit of the quotient. The remainder is obtained by multiplying the final figure by 3 and adding it to the last digit of the quotient. Rule IV

We will illustrate this by an example.
Let us divide 1325 by 7.
7]1 3 2/5[

$$Q = \begin{array}{ccc} 100 & 10 & unit \\ \downarrow & \downarrow & \downarrow \\ 1 & [1\times3 + 3] & [6\times3 + 2] \end{array}$$

$$= \quad 1 \qquad 6 \qquad 20$$

So

$$Q = \begin{array}{ccc} 100 & 10 & unit \\ \downarrow & \downarrow & \downarrow \\ 1 & 6 & 20 \end{array}$$

= 180 (By horizontal addition)

R = 65 = 20×3+5; But 65 = 7×9+2

Thus we have to add 9 to interim quotient while 2 would be the remainder.

Final answer Q = 180+9

Q = 189

R = 2

To give one more example:

7]3 12 5 → 7]312/5

$$Q = \begin{array}{ccc} 100 & 10 & unit \\ \downarrow & \downarrow & \downarrow \\ 3 & (3\times3 + 1) & (10\times3 + 2) \\ & & \downarrow \end{array}$$

$$= \quad 3 \qquad 10 \qquad 32$$

$$\text{Remainder} = 32\times3+5$$
$$= 101$$

Let us rewrite Q and R.

As earlier noted :

$$Q = \begin{array}{ccc} 100 & 10 & unit \\ \downarrow & \downarrow & \downarrow \\ 3 & 10 & 32 \end{array}$$

= 432 (By horizontal addition)

R = 101

= 7×14+3

Final answer:
Q = 432+14
= 446
R = 3

This process is cumbersome no doubt; but it has a certain elegance about it. The student should try to master this method. It will make him familiar with the working of 'horizontal addition' which is needed elsewhere also in Vedic mathematics.

Finally, to explain how this method works.

Let us take a four digit number abcd
How do we work out 7]abcd

By our rule IV :

$$Q = \begin{array}{ccc} 100 & 10 & unit \\ \downarrow & \downarrow & \downarrow \\ a & 3a+b & 9a+3b+c \end{array}$$

Remainder = 3[9a+3b+c]+d
= 27a+9b+3c+d

Now let us see how our answer is correct.

$$Q = \begin{array}{ccc} 100 & 10 & unit \\ \downarrow & \downarrow & \downarrow \\ a & 3a+b & 9a+3b+c \end{array}$$

This is nothing but :
Q = 100a+(30a+10b)+(9a+3b+c)

= 139a+13b+c

But our divisor is 7.

Q×7 = 7[139a+13b+c]
= 973a+1b+7c   ..... (I)
Our R = 27a+9b+3c+d   ..... (II)

Hence original number should be :

= Q×7+R
= I + II
= 1,000a+100b+10c+d
= abcd

Thus, our method was correct.

The student should try to properly grasp what has preceded. **Once he really understands its logic and methodology, he will find the whole calculation simple and interesting. Moreover, it will strengthen his understanding of our ten-based number system and how numbers get built up.**

Problems

P^4: Divide by 8. Use of paper allowed.
   i) 2345, ii) 4594, iii) 9811, iv) 1374

P^5: Divide by 7. Paper allowed.
   1347, 2345, 3456, 2314.

We now turn two-digit divisors like 98, 99. Here the method is different from what is followed for divisors like 7, 8, and 9 as discussed on earlier pages. If the complement of two-digit divisors is small as in 98, 99 many problems can be solved mentally by merely observing the dividend as per Vilokanam.
For instance :

i)	99] 1134	→	Q = 11
			R = 45
ii)	98] 1234	→	Q = 12
			R = 58
iii)	97] 1311[	→	Q = 13
			R = 50

All these solutions have been written down instantaneously. As anybody can see, this is impossible by conventional computation. We explain below how the various problems given above have been solved by Vilokanam (by mere observation).

   i)   99] 1134[

79

Since divisor has two digits, we have to separate the last two digits as they become a part of the remainder.

99]1134   →   99]11/34

Interim remainder = 34.
We take down 11. It becomes the interim quotient. It needs to be added to 34. Hence
99]1134[Q = 11
                  R = 45
ii)   98]1234   →   98]12/34

We take down 12. This is the interim quotient. How to get remainder? Here we note the complement of 98 is 2. We have to multiply the quotient 12 by 2.
So the remainder becomes : $34 + 12 \times 2 = 58$

Final answer :
98]12/34
Q = 12, R = 58

   iii)   Similarly,

97]1311[
Here Q = 13
R = 11+13×3
= 50

This method can be summed as follows: (for two-digit divisors),

I)   **Separate the last two digits of the dividend. It becomes the interim remainder ..... R^1.   ..... Rule I**
II)   **Excluding the last two digits, the remaining digits become the interim quotient, Q^1.   ..... Rule II**
III)   **Multiply the interim quotient by the complement of the divisor. This becomes our new dividend.**
                                        ..... Rule III

IV) **Repeat the process on the new dividend. Its last two digits become the interim remainder R^2 and earlier digits the new quotient, Q^2.**

V) **Continue the process until all the digits are exhausted and the quotient is reduced to zero.**

While these rules may seem complicated, they can be easily applied and enable us to carry out the division mentally without any effort. We illustrate them by solving following three problems.

99]1159[

Interim remainder R^1
= 59 (last two digits)   [Rule I]
Q^1 = Interim quotient = 11 [First two digits]   ..... Rule II

New Dividend = 11 × 1 = 11 [1 is the complement = 100-99]
                                                       ..... Rule III

99]11 → $Q^2 = 0$, $R^2 = 11$ [Rule IV]
Final remainder = 59 + 11 = 70
= $R^1 + R^2$
Final quotient : 11
= $Q^1 + Q^2$

98]2345[

Interim remainder = 45 R^1 ..... (Rule I)
Interims Quotient = 23 Q1   ..... (Rule II)

New dividend
= 23×2 [2 being the complement]
= 46   (Rule III)

98]46 → $Q^2 = 0$, $R^2 = 46$ (Rule V)
Note at the last stage for a two digit divisor the dividend will have only two digits. They automatically become the remainder and the quotient is reduced to zero.

Therefore,
Final quotient = $Q^1 + Q^2$
= 23

Final reminder
= $R^1 + R^2$
= $R^1 + R^2$
= 91

98]12,311[

Interim remainder' = 11   ..... Rule I
Interim quotient' = 123   ..... Rule II

Now dividend = 123×2
= 246   ..... Rule III

We have to divide 246 by 98.
98]246[$\rightarrow$ 98]2/46   (Rule IV)

This leads to
$Q^2 = 2$
$R^2 = 46$

But the process is not complete. We have to carry it out until all digits are exhausted and we get zero quotient.   ..... Rule V

By our rule no. III the new dividend becomes
$4 = Q^2 \times 2$ (2 is complement) for 98.

Now 98]4[$\rightarrow Q^3 = 0, R^3 = 4$

Final quotient : $Q^1 + Q^2 = 123+2 = 125$

Final reminder = $R^1 + R^2 + R^3 = 11+46+4 = 61$

We will now illustrate how by employing the same Vedic technique even the division of 6-digits numbers can be carried out in a smooth way. Consider,

99]11 11 11[

Here $R^1 = 11$ [Rule I]
       $Q^1 = 1111$ [Rule II]

New Dividend = 1111 [Complement is 1] [Rule III]
99]11/11
So
$Q^2 = 11$ [Rule II]
$R^2 = 11$ [Rule I]

Our new dividend will become : 11 [$= Q^2 \times 1$, 1 is the complement]   [Rule III]
99]11 $\rightarrow Q^3 = 0, R^3 = 11$

Final answer :
$Q = Q^1 + Q^2 + Q^3$
   $= 11\ 11 + 11 + 0 = 1122$
$R = R^1 + R^2 + R^3 = 11 + 11 + 11 = 33$

Once the method is grasped the student can perform this operation of a glance.
Even if the dividend has 8 or 10 digits.

To give another example of division involving 6-digit dividend :
98]222222[

$R^1 = 22$ [Rule I]
$Q^1 = 2222$ [Rule II]

Now Dividend = 4444 = 2222×2 [Rule III]
(2 is the complement)
$Q^2 = 44$ Rule II
$R^2 = 44$ Rule I

New Dividend = $Q^2 \times 2$
$= 44 \times 2$
$= 88$ (Rule III)

98]88   [Rule IV]

$\rightarrow Q^3 = 0, R^3 = 88$

Thus
$R = R^1 + R^2 + R^3$
$\quad = 22 + 44 + 88$
$\quad = 154$

But $154 = 98 \times 1 + 56$

So the final answer : $Q^1 + Q^2 + Q^3 + 1$

$Q = 2222 + 44 + 1$
$= 2267$
$R = 56$

This process is perhaps a little complicated for 6 digit dividend.
But for a 4 digit dividend it yields the answer immediately.

For example:
97]1134[

What do we do? Just look at the number.
Last two digits = 34 = R1 [Rule I]
First two digits = 11 = Q1
Complement = 3
New Dividend = 33 [Rule III]

Therefore,
$Q^2 = 0$
$R^2 = 33$ [Rule IV]

$R = 33+34 = R2 + R1$
$= 67$
$Q = 11$

Surely all this can be done rapidly without using paper.

Solve the following mentally.
P⁶:
i) 99]3745,   ii) 99]1136,   iii) 99]7511
iv) 98]2234   v) 98]1234
vi) 97]1211   vii) 97]1511   viii) 97]1622

To illustrate now the methodology and to demonstrate how it works we will take a practical example :

Suppose we want to find :
98] 333333

As per our rules: The last two digits
$= R^1 = 33$ [Rule I]

The remaining digits become Q^1
$= 3333$

Complement $= 2 (= 100-98)$

New Dividend : $3333 \times 2 = 6666$ (Rule III).
Now to understand what lies behind this calculation :

First note : $33,33,33 = 33,33,00 + 33$

$= 333300 + R^1$

If our divisor were 100, the quotient would have been 3333. But it is 98. Obviously the quotient would be bigger than 3333. The question is how much bigger. This is what the entire calculation is about.

Following equations will help us in this regard.

$98] 333333 = 98] 333300 + R^1$

We focus on the first part:
98] 333300   ..... A

This is nothing but

= 3333 + 98] 3333 × 2 or 98]6666   ..... B

To prove now A = B
For this purpose we multiply both sides by 98.

A×98 = 333300
                    I   II
B×98 = 98×3333+6666
I = [100-2] × 3333
= 333300 - 6666

I+II = 33,33,00

Thus A × 98 = B × 98
Hence A = B
Now to review what this means :
          x   y
B = 3333 + 98]6666.

We see x = Q^1 (Rule II)
              y = New dividend = $Q^1 × 2$ (Rule III)

Thus
A = 98]333300   Original dividend
        Quotient Q^1
= 3333 + 98]6666
                              New dividend
We reapply the same process to 98]6666   [Rule IV]

Repeating it we get :
R^2 = 66
Q^2 = 66

New Dividend = 66×2 or $Q^2×2$
= 132 [Rule IV]

98]132

So $\qquad R^3 = 32$

$\qquad\qquad Q^3 = 1$

Next dividend $Q^3 \times 1 = 2 = R^4$

Here the quotient will be obviously zero as the dividend has only one digit. It has to be the remainder. Thus the process has been carried out until the quotient is reduced to zero. $\qquad$ (Rule V)

To compute the final answer:

$Q = Q^1 + Q^2 + Q^3$

$= 3333 + 66 + 1$

$= 3400$

$R = R^1 + R^2 + R^3 + R^4$

$= 33 + 66 + 32 + 2$

$= 133$

$= 98 \times 1 + 35$

Final answer becomes :

$Q = 3401$

$R = 35$

Once the student gets familiar with this process it will work out smoothly.

While applying these steps always keep in mind the logic behind them. This is especially important when you multiply the quotient by the complement to get the new dividend. You should ask yourself : Why am I doing so? Once this is properly grasped it will happen naturally and it won't be something to be memorized and applied by rote. Ideally memory should play no part in mathematics. Always keep in mind why a particular calculation is being made in a certain way, and this will enable you to carry it out without memorizing anything.

We have worked out below some problems by the Vedic method explained above. The answers can be obtained very quickly. By the traditional way it would have been most cumbersome to solve them. Calculations of this type may not prove possible even by the calculators in mobiles as they cannot handle big numbers. Thus in Vedic computation our motto should be **'Beat the calculator'**. To illustrate now the elegance of the Vedic method of division.

i)   99] 11, 11, 11, 11   Q = 11, 22, 33
             4 pairs   R = 44

ii)   99] 11, 11, 11, 11, 11   Q = 11, 22, 33, 44
             5 pairs   R = 55

iii)   99] 11, 11, 11, 11, 11, 11   R = 66
             6 pairs   Q = 11, 22, 33, 44, 55

The student should note the pattern of numbers occurring in the answers. There is much symmetry in them. The student should try to grasp how such numbers get built up in a sequential way.

Once the steps involved in the calculation are properly understood the answers can be written just by looking at the dividends. The student should remember that the dividend at each step gets reduced by 2 digits. The quotients and remainders are similar. One can add them by a simple mental calculation. Without any rough work following three seemingly complex problems can be solved.

99] 11, 11, 11, 11, 11, 11, 11   R = 77
            7 pairs   Q = 11,22,33,44,55,66

99] 11, 11, 11, 11, 11, 11, 11, 11   R = 88
            8 pairs   Q=11,22,33,44,55,66,77

99] 11, 11, 11, 11, 11, 11, 11, 11, 11   R = 0
                                     Q = 11, 22, 33, 44, 55, 66, 77, 89

            9 pairs of one

In each of these problems the remainder is always 11 at every stage. In the last problem for example there are 9 dividends and so 9 remainders which add up to 99. But 99 = 99 × 1 + 0. So the final remainder becomes zero and we add "1" to the interim quotient. **As regards the quotient it consists of 8 rows having 1 to 8 pairs of "11". We can easily work out their addition mentally. The student should try to visualize such a grid in his mind and hold it firmly there till the calculation is complete. He should be able to see the numbers as clearly as he sees them written on a paper in front of him.**

P^7:
Solve using paper:
i) 97]12345,   ii) 95]11167, iii) 98]33333, iv) 97]52345

P^8:
Solve mentally :
i) 99] 3745,   ii) 99] 1134,   iii) 98] 2234,   iv) 98] 1234,
v) 97] 1211,   vi) 97] 1511

Finally, we turn to 3 digit divisors. Here again exactly the same principles apply. We take out initially the last three digits which become the remainder.

The students should note that in case of a single digit divisor, the last digit of the dividend becomes the remainder, in case of two digit divisors, the last two digits become the remainder. In case of a three-digit divisor, it is the last 3 digits which count as the remainder. This is how one finds symmetry and uniformity in mathematics. There are no exceptions.

The calculation becomes easy when the divisor is close to the base number i.e.1000. We have solved in what follows three problems to illustrate how the division by 3 digit divisors works.

I]   To find : 999] 100,001
Here complement = 1 = 1000 - 999.

$R^1 = 001 = 1$ [last three digits]
$Q^1 = 100$ [First three remaining digits]
New dividend = 100×1 (complement) = 100
$Q^2 = 0, R^2 = 100$
Answer $Q = 100, R = 101$

II]   997] 10235
Complement = 3
(Last three digits)
$R^1 = 235, Q^1 = 10$ (remaining digits)
New dividend = $10 \times 3 = 30$
$Q^2 = 0, R^2 = 30$
$Q = 10, \qquad R \quad = R^1 + R^2$
$$= 235 + 30 = 265$$

III]   950]95, 985
Complement = $50 = 1000 - 50$
$R^1 = 985 \qquad Q^1 = 95$
New dividend = $95 \times 50 = 4750$
$R^2 = 750, Q^2 = 4$
New dividend = $4 \times 50 = 200$
$Q^3 = 0, R^3 = 200$
$Q = Q^1 + Q^2 = 99$
$R = 985 + 750 + 200 = 1935$

But
$1935 = 1900 + 35 = 950 \times 2 + 35$

Our final answer becomes
$Q = 101$
$R = 35$

This method can be conveniently used when the divisor is close to 1,000 like 990, 998, 999 etc. But it can also be used easily when the divisor is like 497. In such a case we take 500 as the subbase. We illustrate with an example.

497] 3,45, 789

We first extract the last three digits.

$R^1 = 789$

But here first three digits won't automatically become the quotient as before, we have to multiply it by 2 = 1000/500. So our quotient becomes : 345×2 = 690

$Q^1 = 690$

To get the new dividend we have to multiply Q^1 by 3 [500-497].
Note : This is what we have always done.
= 2070

So new dividend : 2070
What is 497]2070

Applying the same process : $R^2 = 070$
$Q^2 = 2×2$ (Multiplying by 2 as here 500 is the subbase)
= 4

New dividend = 4×3 (3 is the complement)
= 12
$R^3 = 12$

We thus get: $Q = Q^1 + Q^2 = 694$
Interim $R = R^1 + R^2 + R^3 = 789 + 70 + 12$
Interim R = 871
= 497×1+374
Thus the final answer Q = 694 + 1 = 695
R = 374

Therefore, 497] 3,45,789
   Q
= 695 + 374 (R)

We will in the end review how this method works. To illustrate this let us consider.

998]789561
        A R^1
Dividend = 789×1000+561

$$Q^1 \times 1000 + R^1$$

So 1000] 789561 = 789 + 561

Thus the last 3 digits become the remainder and the first three digits the quotient. [Rules I and II]

But we have to find
998]789561
We know the first quotient is 789.
Now 998×789
$= [1000-2] \times 789$
$= 789000 - 789 \times 2$

Thus when we take 789 (first three digits) as the quotient and multiply it by 998 we get a deficit of $789 \times 2 = Q^1 \times$ complement (1000-998).

In order to make it up we take as our next dividend:
$Q^1 \times$ complement   [Rule III]

This is what we have been doing.

In this way we get the answer:
998]789561 = Q = 791
                    R = 143

Finally, to review why we multiply by 2 (1000/500) when our subbase is 500 and base 1,000.

Consider :
500] 4567 = 4000 + 567
$4 \times 1000 + 567$
Here last three digits become the remainder = 567. The quotient becomes 8 [4×2]. This is because : $4 \times 1000 = 4 \times 2 \times 500 = 8 \times 500$
This can also be looked at as
$abc \times 1,000 = 2(abc) \times 500$. Thus 2(abc) becomes the quotient.

Therefore we have to multiply by 2 the first digits (excluding the last 3). This gives us the first quotient.

Here the rule can be stated as : Exclude the last three digits. (for a three digit divisor). To get the quotient multiply the first digits by base/sub-base. **.....Rule VI**

To get the new dividend we have to multiply the quotient by the complement as we have been always doing. This is how the process continues.

Finally, in the end we pose two interesting problems. Those who have properly understand what has preceded so far should be able to solve them easily.

P^9:
The problem is : What is :
9999]12345678 ..... (A)
Now we give below another similar problem. What is:
9998]12345678 ..... (B)

But you don't have to solve (A) or (B). You have to only indicate the difference between them. The answer is a 4-digit number which already occurs in the above questions. No proper work required.

Time allowed : 5 seconds. To be solved by Vilokanan. No paper work.

P^{10} :   What is : 9]8,0$\underbrace{00,000,000}_{9 \text{ zeros}}$

What is: 9]8,$\underbrace{000 \ldots 0}_{n \text{ zeros}}$

All problems to be solved by Vedic method.
$P^1 P^2$: To be solved by mentally.
$P^3 P^4 P^5$: Use paper.

P^1:
    i) $888 = Q, 8 = R$
    ii) $777 = Q, 7 = R$
    iii) $66\ 666 = Q, R = 6$

P^2:
    i) $Q = 149, R = 0$
    ii) $Q = 256, R = 7$
    iii) $Q = 79, R = 0$
    iv) $Q = 167, R = 8$
    v) $Q = 236, R = 7$

P^3:
    i) $Q = 1095, R = 1$
    ii) $Q = 1039, R = 5$
    iii) $Q = 837, R = 6$
    iv) $Q = 7772, R = 8$

P^4:
    i) $Q = 293, R = 1$
    ii) $Q = 574, R = 2$
    iii) $Q = 1226, R = 3$
    iv) $Q = 171, R = 6$

For example, take iv) $8]1374 \longrightarrow 15\ 17$
                         Q^2
$IR = 38 = 8 \times 4 + 6.$ Hence $R = 6$
$1517 \longrightarrow 167 = Q^1, Q = Q^1 + Q^2 = 171$

P^5:
    i) $Q = 192, R = 3$
    ii) $Q = 335, R = 0$
    iii) $Q = 493, R = 5$
    iv) $Q = 330, R = 4$

$7]2314 \longrightarrow 2928 \longrightarrow 2\ 11\ 8 \longrightarrow 3\ 1\ 8 + Q^2 = 12 = 330$
$IR = 88 = 7 \times 12 + 4.$ Thus $R = 4$

Solve mentally.

P⁶: i) Q = 37, R = 82
     ii) Q = 11, R = 47
     iii) Q = 75, R = 86
     iv) Q = 22, R = 78
     v) Q = 12, R = 58
     vi) Q = 12, R = 47
     vii) Q = 15, R = 56
     viii) Q = 16, R = 70

To be solved using paper.

P⁷: i) Q = 127, R = 26
     ii) Q = 117, R = 52
     iii) Q = 340, R = 13
     iv) Q = 539, R = 62

Solve mentally.

P⁸: i) Q = 37, R = 82
     ii) Q = 11, R = 45
     iii) Q = 22, R = 78
     iv) Q = 12, R = 58
     v) Q = 12, R = 47
     vi) Q = 15, R = 56

P⁹: The problem is:
     9999]12345678   . . . . . (A)
     9998]12345678   . . . . . (B)

We have to indicate the difference between the solutions of (A) and (B) in five seconds. How is this to be done?

The answer is: The quotient is the same for (A) and (B). There is a difference in the remainders only. It is 1234.

How has this been worked out?

For A: Since the divisor has 4 digits, we have to set aside the last 4 digits of the dividend. These becomes R1 as earlier. The first four digits then become the quotient. Therefore the quotient becomes 1234.

This is what we have been doing for two-digit, three-digit divisors. The quotient is obtained by excluding the last two/three digits. Hence in case of (A) excluding the last 4 digits, we get the quotient of 1234. In case of (B) too the quotient similarly becomes 1234.

Thus there is no difference in quotients for (A) and (B). Turning now to the remainders:

$$R^1 \qquad R^2$$

For (A) : R = 5678 + 1234

$$R^1 \qquad R^2$$

For (B) : R = 5678 + 2 × 1234

For (B) the complement is 2 against the base of 10,000.

We don't have to calculate R in case of (A) and (B). We only want to know the difference between them. **Thus only by looking at these numbers we can infer that the difference between the two remainders is 1234.**

This is the beauty of Vedic computation. To solve this problem by the traditional method would have required much time and effort; but by the Vedic way merely glancing at these numbers is enough.

P^{10}:   (i)   888,888,888 = Q

9 times

8 = R

Also observe dividend in (1) has 10 digits. The quotient has 9 digits.

This is in order.

$$P^{10}: \quad \text{(ii)} \quad \underbrace{888, \ldots \ldots \ldots}_{n \text{ times}} = Q$$
$$8 = R$$

This kind of generalization easily proves possible in Vedic mathematics. A normal calculator of course will fail at (i) itself. Even by the traditional method to infer such a generalized solution as in (ii) would be difficult.

Section IX
The Digital Roots of Numbers - The Magical 9

Looking at the title of this section, no doubt two questions would arise in the student's mind:

   i)   **What is the digital root of a number?**
   ii)   **Why have we described '9' as a magical number?**

Answers to both these queries are somewhat intertwined. Here we will only say at the beginning that the student is already acquainted with some wonderful features of "9".

To give an instance: we saw in the last section VIII that division by 9 even of long numbers can be carried out quickly and that too by mere observation. For instance, what is:

9]111 111 111

      9 times

We immediately get the answer Q = 12345679. This won't obviously be possible by the traditional method and a calculator probably would be unable to cope with a dividend having 9 digits; but such an instantaneous division can be carried out by the Vedic methodology only when 9 is the divisor. This is why we said earlier that number 9 indeed possesses some special features, and these were brought into prominence by Vedic mathematicians.

The special characteristics of 9 become all the more apparent when we deal with digital roots of numbers. **We need to emphasize here that the concept of digital roots of numbers, as well as of number "9" possessing special features, is found only in Vedic mathematics.**

But what is the digital root of a number? This is defined as the sum of its digits. Carried out in succession if necessary

until it is reduced to a single digit. We give below some instances.

Number   Digital Root
157   1+5+7 = 13 → 4

Table A

	Number	Digital Root		
i)	157	1+5+7 = 13	→	4
ii)	199	1+9+9 = 19	→	1
iii)	257	2+5+7 = 14	→	5
iv)	393	3+9+3 = 15	→	6
v)	192	1+9+2 = 12	→	3

Look at the digital roots of (ii) 199, (iv) 393 and (v) 192. They share an interesting feature. **Suppose we substitute in them 9 by 0; the digital roots would still remain unaffected. Let us verify this for ourselves**.

Table B

	Number	By substituting 9 by zero	Digital Root
ii)	199	100	1
iv)	393	303	6
v)	192	102	3

The student will readily observe that the digital roots of numbers of (ii), (iv) and (v) are identical both in Table A and Table B.

It is to be noted that the digital root of a number remains unaffected by substituting 9 with zero. This is truly a magical facet of 9. ..... **Rule I**

In order that the student thoroughly grasps this concept, we give a few more examples:

	Numbers	Substituting 9 by 0
i)	195	105
ii)	179	170
iii)	189	180

It is to be observed that digital roots of 195/105, 179/170, 189/180 are identical being 6, 8 and 9.

We are now in a position to enumerate some other almost unbelievable properties of 9.

The digital root remains unchanged by adding 9 to itself any number of times ..... Rule II

For example: Digital roots of 9, 18, 27, 36, 99999, etc. will always be the same i.e. 9. This will become evident from the following:

Number	Digital Root
9	9
$9 \times 2 = 18$	$1 + 8 = 9$
$9 \times 3 = 27$	$2 + 7 = 9$
999999	$9 + 9 + 9 + 9 + 9 + 9 = 54 \rightarrow 9$

This is why it makes no difference to the digital root if 9 is substituted by 0. Thus, $99999 \rightarrow 900000 \rightarrow 9 \rightarrow 0$.

Moreover, if any number is added to 9 or its multiples, the digital root remains unchanged. .... Rule III

Thus,   Table A   Digital Root

1+9	$\rightarrow$	10	$\rightarrow$	1
2+9	$\rightarrow$	11	$\rightarrow$	2
3+9	$\rightarrow$	12	$\rightarrow$	3
4+9	$\rightarrow$	13	$\rightarrow$	4
5+9	$\rightarrow$	14	$\rightarrow$	5
6+9	$\rightarrow$	15	$\rightarrow$	6

7+9	$\rightarrow$	16	$\rightarrow$	7
8+9	$\rightarrow$	17	$\rightarrow$	8
9+9	$\rightarrow$	18	$\rightarrow$	9

We have added to 9 various numbers from 1 to 9. But the digital root of the sum is always the same as the digital root of the number added. For example 3 (digital root 3) + 9 = 12 (digital root 3). This aspect is clearly observed from table A given earlier.

It is to be noted that the digital roots of numbers move in a circle from 1 to 9. The following configuration will illustrate this. We have shown digital roots closest to the circular diagram.

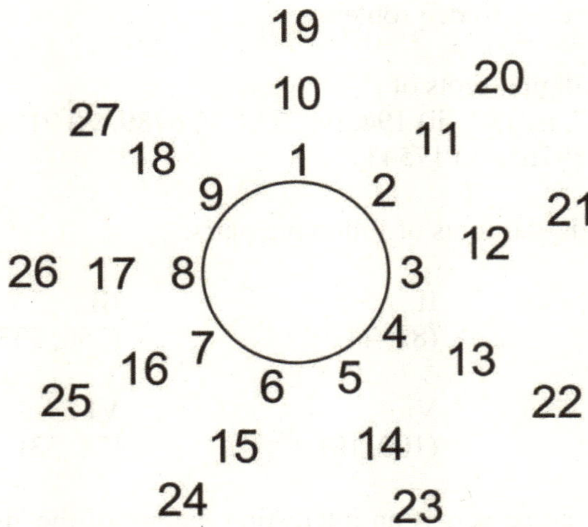

We can thus conclude: **As far as digital roots are concerned, addition of multiplies of 9 to a number makes no difference. It is as if we are adding "0" to that number. .... Rule IV**

This can be stated generally in terms of algebra as:

Digital root of x + 9a (any multiple of 9)
= Digital root of x.

This is because as we saw earlier the digital root of any multiple of 9 is 9 only (=0). [Rule II]

That 9 becomes equivalent to "0" while computing digital roots is expressed in the following Vedic sutra elegantly.

शून्यं साम्यसमुच्चये

This means "when the total is the same, it is naught".

शून्यं = Zero
साम्य = Equal
समुच्चये = Total

Given here are some problems:

P[1]: Find digital roots of :
  i) 157, ii) 189, iii) 194, iv) 2111, v) 5789, vi) 9119, vii) 240, viii) 89765, ix) 17543

Find the digital roots of following pairs:

I	II	III
(13,22)	(89, 44)	(756, 333)

IV	V	VI
(495, 630)	(100, 10)	(78, 33)

Here, we come across an interesting feature of the 'digital roots' as explained below.

First the student should note that numbers in each pair are different.

But digital roots of each pair are the same.

Table C

Pairs	Digital Root
I	4
II	8
III	9
IV	9
V	1
VI	6

What conclusion can we draw from the above table C? The obvious conclusion is:

The digital root can be the same even if the two numbers are widely different.

We can thus affirm:
The sameness of digital roots in no way implies sameness of the numbers. On the other hand, if the two numbers are the same, their digital roots have to be one and the same.
.... Rule V

Now we will see how the above principle of digital roots finds practical application.

Suppose we have to add any two big number, like
I   II
15,791 + 16,999.

                                        III
Presume a student writes the sum as : 32,791.

Is this a correct answer? This can easily be checked by finding their digital roots and adding them.

Digital root of I = 5
Digital root of II = 7
Digital root of I + II = 3
Digital root of III = 4

Thus the sum is certainly incorrect. The correct answer is 32,790.

Digital root = 3

Thus after a particular computation is carried out then on the basis of digital roots, we can quickly ascertain whether the answer is correct or wrong. But here following should always be kept in mind.

If the digital root of the sum differs from the sum of digital roots of the two numbers, then the computation is certainly wrong. But if they agree, no inference can be drawn as regards correctness of the answer   ..... Rule V

This is simply because as explained earlier two unequal numbers can very well have the same digital root.

In mathematics this will be rigorously expressed as "Equality of digital roots is a necessary condition for two numbers to be equal but it is not sufficient."

We will now illustrate how to utilize this concept to check the correctness of computations. We start with additions.

The digital root of the sum of two numbers is equal to the sum of their digital roots. For example : take 114 and 115.
Digital root of 114 = 6
Digital roots of 115 = 7
Sum of digital roots = 6+7 = 13 $\rightarrow$ 4.
But 114 + 115 = 229
Digital root of 229 = 13 $\rightarrow$ 4.

This will also apply to subtractions and multiplications. To illustrate :
I   II
2199   - 222
   III
= 1977

Digital root of I = 3
Digital root of II = 6
I - II = 3 - 6

Let this not fox the student as 6 cannot be subtracted from 3. How can this be tackled? It is very simple. We know that the digital root remains unaffected by addition of any multiples of 9. We can therefore write 3 as 93 = (9×10+3). Thus,

I - II = 93 - 6
= 87

87 → 6 digital root.

Thus the difference of digital roots of I and II = 6.

Digital root of the subtraction at III = 6.

Thus, digital root of I - II = Digital root of I - Digital root of II.

Finally, we turn to multiplication.
Consider,

    I   II
100,005 × 100,001

Digital root of I = 6
Digital root of II = 2
Product of digital roots of I and II = 6 × 2
→ 3

The product is :
   III
100,00600005

[The student should be able to do this multiplication mentally as explained in an earlier section.]

Digital root of iii = 3
Hence,
Digital root of I × Digital root of II
= Digital root of the product.

Finally, to explain how this comes about. **The "digital root" can be looked as the essence of a number. This remains unaffected, unchanged even after a computation is carried out. Now to prove this proposition:-**

Let us take any two numbers, x and y. We will establish here the followings:

Why is it :
i)   Digital root of x + Digital root of y = Digital root of x + y.
ii) Digital root of x - Digital root of y = Digital root of x - y.
iii) Digital root of (x) x Digital root of (y) = Digital root of xy.

We will first prove (i). The proofs for (ii) and (iii) are very much similar.

Any number x can be written as :
x = 9a + b
For instance : Take 192
Here x = 192 = 9 × 21 + 3
So a = 21, b = 3

Now Digital root of x = 9a+b=b [Rule iv]
This corresponds to : Digital root of 192 = 9×21+3=3

Similarly, we split up by y as :
y = 9c + d

In the same way as above the digital root of y = d [Rule iv]

Hence, digital root of x + digital root of y = b + d L

Let us now look at the sum x+y.

It gets split up as :
[9a + b] + [9c + d]
= 9[a+c]+b+d

Hence digital root of x+y = b+d   .... M
L = M

This proves the proposition (i).
 The proof for the proposition (ii) is identical.

Digital root of x - y works out as :
x - y = (9a + b) - (9c + d)
= 9(a - c) + (b - d)

Hence the digital root of x - y = b - d

This is nothing but :
Digital root of x - Digital root of y.

Finally, turning to (x) $\times$ (y)
What is xy

This gets split up as per the earlier notation :
[9a + b] $\times$ [9c + d]

= 81 ac + 9 bc + 9 ad + bd
= 9 [9ac + bc + ad] + bd

In terms of the digital root :
9[9ac + bc + ad] $\rightarrow$ 0   [Rule iv]

Thus the digital root of x.y
= bd

This as can be easily seen is nothing but :
(Digital root of x) $\times$ (Digital root of y).
Thus all the three propositions given at the beginning are proved.
We now turn their practical application.

———

P^2: Find by computing digital roots which of the following propositions are certainly wrong :
i) $97 \times 98 = 9507$
ii) $93 \times 91 = 8460$
iii) $101 \times 105 = 10605$
iv) $1134 - 91 = 1042$
v) $575 - 22 = 554$
vi) $9578 + 1234 = 10,812$

Solutions

Section XI

Problems P^1 :   i) 4 ii) 9 iii) 5 iv) 5 v) 2 vi) 2 vii) 6 viii) 8 ix) 2

Problems P^2 :   i) Wrong ii) Wrong iii) Correct iv) Wrong
v) Wrong vi) Correct

We illustrate by solving the first one, P^2(i)

Digital root of 97 = 7

Digital root of 98 = 8

Product of digital roots = 56 $\rightarrow$ 2

Digital root of 9507 = 3

Hence i) is wrong.

Other problems to be solved similarly. In the case of (ii), (iv) and (v) the digital roots of two sides don't agree. Hence the answers are certainly wrong.

As regards (iii) and (vi) the digital roots agree. But we cannot infer from it the correctness of the answer. Incidentally both the answers are correct !

Section X
How to cube a two-digit number through addition: the Vedic way.

The Vedic method of cubing two-digit numbers is especially useful when both the digits are small like 31, 41, 12 etc.

This method can easily be understood in terms of the following algebraic formula:

$$[a+b]^3 = a^3 + 3a^2b + 3ab^2 + b^3 \qquad \ldots\ldots\ldots (A)$$

Let us try to cube 31 as per the above formulation. In 31, $a = 3$ and $b = 1$; We write 31 as $3\times10+1$.

We do not cube the "ten" in the above split-up. **This is because the positioning of the digits ensures that the "10" factor is fully accounted for**. We will see later how this happens:

$$31^3 = \begin{array}{cccc} 3^3 & 3\times 3^2\times 1 & 3\times 3\times 1^2 & 1^3 \\ \downarrow & \downarrow & \downarrow & \downarrow \\ a^3 & 3a^2b & 3ab^2 & b^3 \end{array} \qquad \ldots\ldots (B)$$

The numbers at B are nothing but :
$$31^3 = 27,\ 27,\ 9,\ 1 \qquad \ldots\ldots (C)$$

To get the answer we have to carry out what is called the "horizontal addition". We had explained this process earlier. Thus,

$$31^3 = 27\ \ 27\ \ 9,\ 1$$

$$= 29\ 7\ 9\ 1$$

$$\downarrow \qquad \text{By horizontal addition.}$$

$$= 29{,}791 \qquad (\overline{C})$$

Surely, this is far quicker than direct multiplication.

While we have explained the method of horizontal summation from right to left as carried out in this case in an earlier section, we will quickly review it here.. Rewriting (C) :

$$
\begin{array}{cccc}
(i) & (ii) & (iii) & (iv) \\
27 & 27 & 9 & 1 \\
\downarrow & \downarrow & \downarrow & \downarrow \\
a^3 & 3a^2b & 3ab^2 & b^3
\end{array}
$$

$31^3 = $ .....(D)

It should be noted that the figures from (i) to (iv) are in the position of 1,000, 100, 10 and unit.

$$
\begin{array}{rcr}
\text{Therefore D} & = & 27,000 \\
& + & 2,700 \\
& + & 90 \\
& + & 1 \\
\hline
& & 29,791
\end{array}
$$
Note $(\overline{C}) = D$

27,000 corresponds to 30^3, 2700 $= 3\times(30)^2$, 90 $= 3\times30$. We would get the same set of figures if we were cubing $[30+3]^3$ as per algebraic formulation $(a+b)^3 = a^3+3a^2b + 3ab^2 + b^3$ given on the prepage. We took earlier "a" as "3" in our calculation only to facilitate the computation. **Thus though we did not show the zeros explicitly in our calculation they automatically get reflected due to positioning of the digits.**

We can now easily understand the rationale behind horizontal addition from right to left.

Let us take a look again at the figures in D. They go as

$$
\begin{array}{cccc}
(i) & (ii) & (iii) & (iv) \\
1000 & 100 & 10 & unit \\
\downarrow & \downarrow & & \\
27 & 27 & 9 & 1
\end{array}
$$

Now 27 at (ii) = 2,700 = 2,000 + 700

So we transfer 2,000 to the left which is a 1,000 position. To do this we have to merely add "2" to 27 which is in 1000th position. Thus 27 becomes 29. The remaining 700 gets represented by "7" being in the hundredth position. Thus:

27, 27, 9, 1
$\rightarrow$ 29,791.

This brings out the rationale of the 'horizontal addition'.
We will illustrate it further with two more example. Let us consider 43^3.

Here as per the notation used earlier, a = 4, b = 3. We compute $43^3 = [40+3]^3$ in what follows as previously done.

$$43^3 = \begin{array}{cccc} 4^3 & 3(4^2 \times 3) & 3(4 \times 3^2) & 3^3 \\ \downarrow & \downarrow & \downarrow & \downarrow \\ a^3 & 3a^2b & 3ab^2 & b^3 \end{array}$$

These figures become

$\rightarrow$ 64   144   108   27

Adding from right to left, we get :

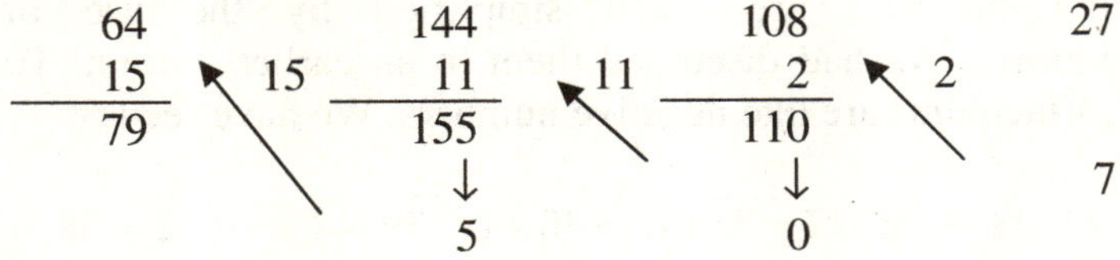

Thus, the final answer becomes :
79,507

The student should remember that the horizontal addition is always carried out from right to left. Therefore the above chart should be read from right to left. Since the digits are on the higher side, the above may appear a little cumbersome. The student should go over this problem again and again so that he can carry out the 'horizontal addition' with confidence and ease.

In our next illustration we will work with smaller digits. This will bring home to the reader how the above method is quite good when the number to be cubed is close to 10, 20, 30 etc. Let us work out, 51^3:

As per the formula $a^3 + 3a^2b + 3ab^2 + b^3$, this will become [a=5, b=1]

$$51^3 \quad = \quad 5^3 \quad 3\times5^2\times1 \quad 3\times5\times1^2 \quad 1^3$$

$$= \quad 125 \quad 75 \quad\quad 15 \quad\quad 1$$

This becomes :

$$125 \quad\quad\quad 75 \quad\quad\quad 15 \quad\quad\quad 1$$

$$\nwarrow 7 \quad\quad\quad \nwarrow 1$$

= 1,32,651

So the answer is : 1,32,651

It would be thus seen that through the device of 'horizontal division', the process of "cubing" in the Vedic system becomes far quicker, simpler and error free as you deal with smaller numbers.

Finally, we will see how such calculations involving cubing of two-digit numbers are much simplified by the use of "Vinculums". We had discussed them in an earlier section. To recall, Vinculums are like negative numbers. We have seen :

$$39 \rightarrow 4\bar{1}, 38 \rightarrow 4\bar{2}, 27 = 3\bar{3} \; [4\bar{1} = 40 - 1 = 39, 4\bar{2} = 40 - 2 = 38$$
$$3\bar{3} = 30 - 3 = 27]$$

The main features of Vinculum are :
$$\bar{1} \times \bar{1} = 1, \bar{1} \times 1 = \bar{1}$$

Let us see how we can comfortably cube a number like 39 using Vinculums.

$$39 \rightarrow 4\bar{1}$$

$39^3 = 4\overline{1}^3$

As per the formula $(a+b)^3 =$
$a^3 + 3a^2b + 3ab^2 + b^3$ [$a = 4$, $b = \overline{1}$]
$4\overline{1}^3 = 64 \qquad 3\times4^2\times\overline{1} \qquad\qquad 3\times4\times(\overline{1} \times \overline{1})$
$(\overline{1}^3)$

$= 64 \qquad \overline{48} \qquad 12 \quad \overline{1}$

$= 60 \qquad \overline{7} \qquad 2 \qquad \overline{1}$

But $60\overline{7} = 593$
$2\overline{1} = 19$
Hence the answer is
$39^3 = 59{,}319.$

The student should thoroughly familiarize himself with Vinculums. To give one more example:

$29^3 = 3\overline{1}^3$

$3\overline{1}^3 = 27 \qquad 3\times3^2\times\overline{1}, \; 3\times3\times\overline{1}^2 \qquad \overline{1}^3$

$= 27 \qquad \overline{27} \qquad 9 \qquad \overline{1}$

$= 27 \qquad \overline{27} \qquad 9 \qquad \overline{1}$

$= 25 \qquad \overline{7} \qquad 9 \qquad \overline{1}$

But $\qquad 25\overline{7} = 243$
$\qquad\quad 9\,\overline{1} = 89$

Thus, the answer :
$29^3 = 24{,}389$

The operation of Vinculum may appear a little confusing to the student. But there is no cause for confusion. We had promised at the very beginning in the foreword itself that there is no mystery in mathematics. Everything should be crystal clear as sunlight. We have therefore elaborated a little in what follows the above computations with Vinculums.

Let us take a look again at: 29^3

	i)	ii)	iii)	iv)	
=	27	$\overline{27}$	9	$\overline{1}$	X
=	27	$\overline{7}$	9	$\overline{1}$	.....Y
=	25	$\overline{7}$	9	$\overline{1}$	

$= 25\overline{7}9\overline{1} = 24389$

Let us see how horizontal addition also works with Vinculums.

$$X = \quad \begin{matrix} (i) & (ii) & (iii) & (iv) \\ 27 & \overline{27} & 9 & \overline{1} \end{matrix}$$

(i)   This is at 1000th position and
(ii) This is at 100th. So these numbers really are : 27,000 and -2700

But   27,000 - 2700 = 27,000 - 2,000 - 700 =
      = 25,000 - 700 = 24,300   ..... (Z)

By horizontal addition : $27 \quad \overline{27}$
$$25 \quad \overline{7}$$

[When we transfer $\overline{2}$ to left we actually deduct 2,000 as done in (Z) above]

$$25\overline{7} = 243 \qquad\qquad ..... \text{V}$$

V really stands for 24,300 as 24 is at 1000th position and 3 at the 100th.

But this is same as (Z).

Consider now (iii) and (iv) or 9 and $\overline{7}$ which are at 10th and unit position.

$9\overline{7} = 89 = 90 - 1 ..... \text{VI}$

Combining V and VI we get the final answer = 24,300+89=24,389

P^1:   Solve the following by the Vedic method :
        i) 21^3, ii) 42^3, iii) 52^3, iv) 61^3

P^2:   Do mentally :
        i) 11^3, ii) 12^3, iii) 13^3, iv) 14^3, v) 21^3, vi) 22^3

Hint: Write the answer digit by digit from the right. Close your eyes and concentrate. This will help to inculcate the habit of concentration in you.

P^3:   Solve using Vinculums
        i) 19^3, ii) 28^3, iii) 37^3, iv) 47^3

Finally, the student may wonder whether there is any difference between 'horizontal addition' and 'vertical addition' which we normally do. There is in reality no difference.

The vertical summation involves "carry over". In the "horizontal summation" there is no carry over as such directly. But that is what we do indirectly.

Take the following set of numbers

1000	100	10	1
15	10	17	21

Written as : 15 10 17 21

They becomes by horizontal addition

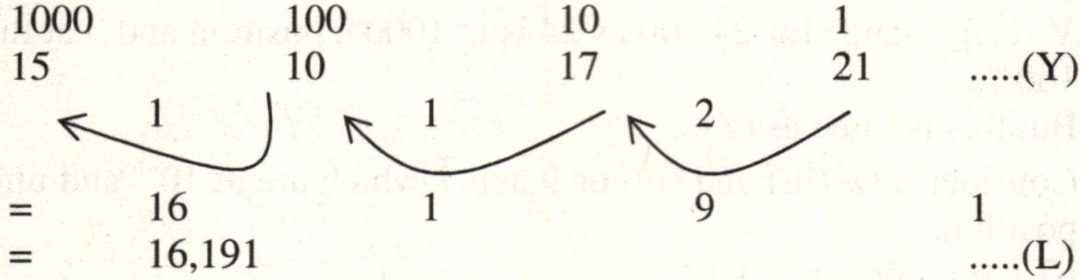

1000	100	10	1	
15	10	17	21	.....(Y)

=   16   1   9   1

=   16,191   .....(L)

Here "2" is carried over from 1st to 10th and "1" is carried over from 10th to 100th position as well as from 100th to 1,000th.

Now let us add them vertically.

1000	100	10	1
15	10	17	21

They stand for :

=   15,000

        1,000   1 added to five

+   170   1 added to zero

          21   2 added to 7

—————————

      16,191

It will be easily noted this is exactly what we have done at (Y) while carrying out the horizontal addition.

.....(M)

L = M

The horizontal addition works due to the decimal system of our numbers. It says a digit's value depends upon its position in the number. For instance, in 1111 the value of "1" goes up from 1 to 10 to 100 to 1,000. Thus when we say 21 is in a unit position it stands for 20+1. During horizontal addition only 1 remains in the unit position while 2 moves to the 10th position which automatically gives it the value of 20.

In the same way the process of vertical summation also is based on our famed decimal system. When we carry over "1", "2" .... "5" etc. from one column to another we actually transfer 100, 2000, etc. depending on the position of the digit.

Answers to Section X

P^1: $\quad 21^3 = 9261$
$\quad\quad\quad 42^3 = 74088$
$\quad\quad\quad 52^3 = 140608$
$\quad\quad\quad 61^3 = 226981$

P^2: $\quad 11^3 = 1331$
$\quad\quad\quad 12^3 = 1728$
$\quad\quad\quad 13^3 = 2197$
$\quad\quad\quad 14^3 = 2744$
$\quad\quad\quad 21^3 = 9261$
$\quad\quad\quad 22^3 = 10648$

P^3: $\quad 19^3 = 6859$
$\quad\quad\quad 28^3 = 21952$
$\quad\quad\quad 37^3 = 50653$
$\quad\quad\quad 47^3 = 103823$

Section XI
The discovery of infinity - the crowning glory of Vedic mathematics

We know that Vedic seers are given the credit for their discovery of zero. It had revolutionized the number system. We have seen in earlier sections that they had introduced the concept of base numbers like 100, 1000, 10,000, etc., and showed how they can be used to greatly simplify a certain set of numerical computations.

We have seen before how by turning to base numbers one can mentally solve problems like $(1001)^2$, 995×997, 9999×9992, etc., in a jiffy. Vedic seers had devised new methodologies of computation like 'horizontal addition' which greatly facilitate multiplication by substituting it with addition; we saw earlier in depth how by the technique of Vedic division solutions of even seemingly complex problems like $\underbrace{99]11\ 11\ \dots..\ 11,}_{9\ \text{pairs}}$
1234]12345678, etc. could be reached almost instantaneously. This is something which even a computer would fail to do. In fact the Vedic mode of division is undoubtedly highly `elegant and occupies the pride of place in the Vedic methodology of computations.

But perhaps the greatest achievement of Vedic seers was their discovery of the concept of infinity and putting it to practical use. In modern mathematics the concept of infinity was unknown until recent centuries. It was one John Wallis who had in 1657 put it forth.

Contemporaneously, Sir Isaac Newton (1643 - 1727), the great physicist, used it to evolve the methodology of "Calculus" which was to become the very kernel of modern mathematics in subsequent decades.

But Vedic seers had grasped what "infinity" truly connoted far, far earlier. Interestingly enough, ancient Greeks who were also keenly interested in scientific studies were baffled by what "infinity" meant. A number of paradoxes have been attributed to one Zeno of Elea, a well-known Greek thinker. These paradoxes which involved the concept of infinity aimed to establish that many of the daily events like motion could not simply take place.

The paradoxes appeared to be logically consistent. As a result a turmoil was created among Greek thinkers. **The fallacy in them related to a wrong visualization of infinity.**

Let us first see how Zeno of Elea established in a seemingly logical way the impossibility of motion.

Surely this was a totally absurd conclusion. But how did he reach it?

To go from point A to point B, one has to first reach the mid-way point C. See the figure below.

A ... C ... B

But to go from A to C one has to go again to a mid-way point say D.

A ... D ... C

Thus, the journey gets divided into infinite segments. To traverse each segment will take some time. **The sum of infinite time spans will be necessarily infinite, and not finite. Thus the journey from A to B will prove impossible as it would require endless time.**

We will explain shortly the contradiction in this argument. Another paradox of Zeno of Elea which is similar goes as follows:-

It says an object can be either infinitely big or have no dimensions at all. Why was such a claim against all experience made? It was argued :

An object is made up of innumerable infinitesimal parts. Now there are only two possibilities :

   i)   Each infinitesimal part will have some dimension, however small.

   ii)   Or it will have no dimensions whatsoever.

In the first case the totality of infinitesimal parts will necessarily make the object infinitely large. In the case of second possibility if each part has no size, the object as a whole will have no visibility whatever. This is again an absurdity.

If we look at both these paradoxes, they involve a common premise, namely, a "sum of infinite units will necessarily become infinite." But this is false. This is because the sum of infinite numbers can very well be finite as illustrated below.

$$1 = 1/2 + \frac{1}{4} + \frac{1}{8} \ldots\ldots \text{ to infinity}$$

On the R.H.S. there are infinite numbers which are powers of 1/2. But their sum turns out to be nothing more than 1. The R.H.S. is called a geemetric progression which stretches into infinity. How this series involving infinite terms has been summed up will be shown later.

Therefore a journey involving infinite segments can and does take finite time while an object with infinite infinitesimal units can have a finite dimension. In this way both the paradoxes of Zeno of Elea are resolved.

Notably enough, the Vedic seers had very much grasped the crucial fact that an infinite series can have a finite sum. They used it to evolve a simple method to calculate "recurrent decimals" of certain fractions.

But what are recurrent decimals?

Fractions can be divided into two types. Fractions like 1/2, 1/4, 1/5 can easily be expressed in terms of decimals. For example: 1/2 = .5, 1/4 = .25, 1/5 = .2

But many factions are not so simple to express in decimals. For example, 1/3 = .33 .... This stretches into infinity. This is an example of a recurrent decimal.

All fractions where the denominator is a prime, other than 2 or 5, give rise to strings of recurrent decimals.

This is for the simple reason that if the denominator is 2, 5 or their square, multiple, etc., it can divide 10, 100, 1,000 etc. as the case may be. For instance:

$$1/2 = .5, \qquad \frac{1}{4} = \frac{25}{100} = .25, \qquad \frac{1}{8} = .125$$

This happens as 4 divides 100, 8 divides 1000, etc. Similarly, $\frac{1}{5} = .2$, $\frac{1}{25} = .04$, $\frac{1}{20} = .05$

But in case the denominator is a prime other than 2, 5 it cannot possibly divide 10, 100, 1000. This gives rise recurrent strings of decimals. In case of 1/3 the string consists of only one digit "3" and hence we write : 1/3 = .333 ..... on and on.

But such strings usually require tedious computation. For example, let us see how we can express 1/19 in decimal terms.

$$1/19 = \quad 19]100[.0526315789473684$$

$$\begin{array}{r} 95 \\ \hline 50 \\ 38 \\ \hline 120 \\ 114 \\ \hline 60 \end{array}$$

$$\underline{57}$$
$$\underline{30}$$
$$\underline{19}$$
$$\underline{110}$$
$$\underline{95}$$
$$\underline{150}$$
$$\underline{133}$$
$$\underline{170}$$
$$\underline{152}$$
$$\underline{180}$$
$$\underline{171}$$
$$\underline{90}$$
$$\underline{76}$$
$$\underline{140}$$
$$\underline{133}$$
$$\underline{10}$$
$$\underline{57}$$
$$\underline{130}$$
$$\underline{114}$$
$$\underline{160}$$

Continuing $1/19 = \quad 19]160[8421$
$$\underline{152}$$
$$\underline{80}$$
$$\underline{76}$$
$$\underline{40}$$
$$\underline{38}$$
$$\underline{20}$$
$$\underline{19}$$
$$\underline{1}$$

Thus, $1/19 = .052631578947368421$, to infinity   ..... (A)

This is a string consisting of 18 digits. **As noted above the string ended with 1. In this way the string repeats itself.** This is how recurrent decimal strings occur.

But why did we do this tedious calculation?

We did it only to bring home to the reader how tedious and time-consuming it is to work out these strings by the traditional method.

The Vedic mathematicians could devise interestingly a simple technique by which the above computation could be done smoothly. We will first explain this technique and then show how it is derived from the concept of infinity.

What is this technique of Vedic mathematicians like?

First note that when a recurrent string finishes it ends with 1. [see A on the earlier page]. **The end digit has to be "1" - the numerator in 1/19.** This is a the string recurs.

In the Vedic system we start from the right. This is diametrically opposite of the normal method when we work out from left. The student will recall the Vedic division too commences from the right while in the traditional method we begin from the left.

So our digit on the extreme right is 1. We write it as 1; to get the next digit we merely multiply it by 2, which gives us 2. We again repeat the process giving 4; this leads to 8. Then 8 becomes 16. We write it as ^{16}. Our string now consists of

$\rightarrow {}^16 \quad 8 \quad 4 \quad 2 \quad 1$

What do we do with ^{16}? We multiply 6 by 2 (12) and add 1 to it. The next digit becomes 3 with superscript 1. Our string now becomes

$^13 \quad {}^16 \quad 8 \quad 4 \quad 2 \quad 1$

Repeating the process we get
$(3 \times 2)+1 = 7$

This gives us 7. Multiplying by 2 we get ^{14}.

Thus our string now becomes :
$9 \quad {}^14 \quad 7 \quad {}^13 \quad {}^16 \quad 8 \quad 4 \quad 2 \quad 1$

Continuing the process we get :

^{1}0 5 ^{1}2 6 3 ^{1}1 ^{1}5 ^{1}7 ^{1}8 9

Our answer now becomes :
0 5 2 6 3 1 5 7 8 9 4 7 3 6 8 4 2 1   ..... to infinity ..... (B)

It will be readily observed
A = B

The string recurs there is '1' an extreme right. With some practice, the digits in B can be worked out rapidly without any possibility of error.

But the student will ask, or atleast should ask, at this stage why did we multiply each digit by 2 to get successive decimals.

Here is where the concept of infinity comes into play. The Vedic mathematicians knew how to sum up an infinite series. Modern mathematics took at least 2500 years to catch up with them.

But first what is a geometric series ?
A sequence like 2, 4, 8, 16, 32, 64 is called a geometric series. There is a common multiplicant between successive numbers. In the case above it is 2.

Now, 1/19 is a sum of the following infinite progression :

$$1/19 = \frac{1}{20} + \frac{1}{(20)^2} + \frac{1}{(20)^3} \qquad ..... \text{ to infinity}$$

This is as per the formula :
$$S = a + a^2 + a^3 ....$$
$$= \frac{a}{1-a}$$

But here a = 1/20. So the sum of $1/20 + \frac{1}{20^2} + \frac{1}{20^3}$

Works out as

$$\frac{^1/_{20}}{1-^1/_{20}} = \frac{1}{19}$$

Thus,

$$\frac{1}{19} = \frac{1}{20} + \frac{1}{20^2}$$ ..... to infinity

The Vedic mathematicians used "2" as a multiplicant to get digits of the string as the fraction 1/19 is a sum of 1/20 + $\frac{1}{(20)^2}$ **......**

The author is well aware that some of this stuff would be beyond the understanding of students at the elementary level. We have written all this so that the students are convinced that Vedic mathematicians were indeed well aware of the concept of infinity much ahead of their times and could apply it successfully to tackle practical problems.

Finally, we will give one more interesting feature of these strings.
We saw earlier
1/19 = .052631578947368421 .... to infinity

Note there are 18 digits. This is the general rule : the number of digits is one less than the denominator (19).

But how does one compute 18/19. No elaborate calculation is required. We merely have to subtract all the digits in the string of 1/19 from 9 one by one. Thus,
18/19 = .9473684210
52631578

Now let us write these two strings one below another.
18/19 = 9473684210521578   ..... to infinity
1/19 = .052631578947368421   ..... to infinity

Adding the two sides separately
L.H.S. = 1 [18/19 + 1/19]
R.H.S. = $\overline{S}$ = .99 .... to infinity.

But what is : $\overline{S}$ = .9 9 ..... to infinity
we use the formula :
$S = a + ab + ab^2$ ..... to infinity
$= \dfrac{a}{1-b}$

In the above case: a = 9/10
b = 1/10
$\overline{S} = \dfrac{9/10}{1-1/10}$
$= \dfrac{9/10}{9/10}$
= 1
So $\overline{S}$ = 1 = L.H.S.

Thus a sum of infinity numbers is reduced to a single digit.

Finally, we turn to ancient scriptures to show how the concept of infinity gets enunciated in them.

The invocation of the Ishā upanishad says inter-dia:

''पूर्णात्पूर्णमुदच्यते,
पूर्णस्य पूर्णमादाय
पूर्णमेवावशिष्यते''

This means:

"Infinity remains unchanged whether another infinity is added to it or subtracted from it."

This might appear to be an impossibility but this is exactly how our number system operates.

We first give the sequence of natural numbers which stretches to infinity :

1, 2, 3, 4, 5, 6   ..... on and on.

Suppose we remove all even numbers from it. What remains is the sequence of odd numbers stretching into infinity.

It goes as : 1, 3, 5, 7 ....

Thus if infinity (even number) are removed from infinity (all numbers) still infinity remains (odd numbers). Similarly, infinity added to infinity still leads to infinity only. This is seen from the following :

1, 3, 5, 7, 9, 11, ... infinity (odd number)
2, 4, 6, 8, 10   ..... to infinity (even numbers)

Combing the two :
1, 2, 3   ..... infinity (natural numbers)

Thus the above-mentioned Upanishadic invocation stanza is fully borne out.

It is indeed to the credit of Vedic mathematicians that they fully grasped that infinitude and finiteness could harmoniously coexist. This is what is demonstrated in the summation of an infinite geometric series which we have carried out repeatedly earlier.

This interestingly enough finds reflection in the Hindu theology too. We all know about Lord Krishna's universal form (Vishvaroop) extending upto infinity in all directions. But he also manifested himself in a finite, human form. How to do we reconcile the two?

This is nothing but the "infinity" being encompassed within a limited space as happens in the summation of a geometric series.

One finds the same phenomenon in the relationship between Soul and body. We all know Soul is all-pervading; how does it then subsist in a finite body. Moreover, without Soul there would be no body. This is nothing but the coexistence of infinitude and finiteness, both being complementary to each other.

Thus our Vedic seers had truly grasped what infinity was and it became so to say a bond between Vedic mathematics and Hindu metaphysics.

Dear readers,
Finally let us remind ourselves that the edifice of Vedic Mathematics rests on the humble zero. Its discovery along with the 10-based number system was hailed by the famous astrophysicist Jayant Naralikar as an epic one which brought about a 'global revolution' in the world of mathematics.

We end our little tract by expressing the pious hope that the boys and girls going through it will feel inspired to become trailblazers in mathematics as their forebears were.

About the Author

Dr. Prakash Joshi joined the Indian Foreign Service in early 1970s. He went on to hold various senior positions with distinction in different parts of the world, including the Gulf region, Africa, Europe, the Caribbean, etc. While posted to Guyana, South America, as High Commissioner, he delivered on their TV network discourses on the Gita for over two years.

Dr. Prakash Joshi was educated in Cambridge, Gonville and Ceius College, U.K. where he did Tripos in Mathematics. He had earlier graduated with high honours from the University of Bombay specializing in Mathematics. He was then nominated as a National Scholar and was awarded the prestigious Tata Scholarship for higher Education for Indians. During 1980s while posted in Delhi he completed his doctorate through the Jawaharlal Nehru University in the field of International Relations. He also qualified at that time as Interpreter in Arabic through the school of Foreign Languages, New Delhi. Dr. Prakash Joshi possesses excellent knowledge of Sanskrit and has a distinguished family background; his father Dr. V.M. Joshi, I.C.S., D.Sc. was a renowned statistician.

Dr. Prakash Joshi has several publications to his credit in the field of Hindu metaphysics. These include

1) Saga of Hinduism - Volumes I and II : M. D. Publications, New Delhi.
2) Introduction to Sankara's Advaitism : Motilal Banarasidass Publishers, New Delhi. Revised Edition published by Motilal Banarsidass International, Delhi.
3) Path to Liberation : from known to unknown : Motilal Banarasidass Publishers, New Delhi. Revised edition published by Motilal Banarsidass International, Delhi.
4) From Vedanta to Modern Science : Ocean Books Pvt. Ltd., New Delhi
5) Present Day Gita - a way of Life : Publisher K. G. Ganju, New Delhi.
6) The Golden Decade 2005-2015, My life as Hony Consul General : Publisher : K. G. Ganju, New Delhi

Dr. Prakash Joshi is also engaged in systematically translating Sanskrit classics into English. His published works are:
7) Translation of Shakuntalam into English: Publisher : Somanath Sanskrit University.
8) Translation of Mudrā Rakshasam into English: Publisher: Somanath Sanskrit University.
9) Translation of Meghadūtam into English: Publisher : Motilal Banarasidass International, Delhi.

Forthcoming translation by the author of Sanskrit Classics into English are:
10) Mālvikāgnimitram.
11) Vikramorvaśīyam
12) Ṛtusaṁhāra
13) Raghuvaṃśa (First five sargas)

Dr. Prakash Joshi thus hopes to translate into English all the main works of Mahakavi Kalidas.

His forthcoming publications by Motilal Banarasidass International are :
14)   The Hidden treasures of the Gita.
15)   Quest : My spiritual journey - in search of happiness, Volumes I & II .
16)   Translation of Sundarkand into English.
17)   Translation of Kalipuja into English.
18)   Jay Veer Hanuman. English translation of Hanuman Chalisa, Hanuman Ashataka and Hanuman Arti. Original verses with transliteration.

Future publications of the author
19)   A concise commentary on the Gita : original verses with transliteration. Extensive quotations from Gyaneshwari and Sri Sri Paramahansa Yogananda.
20)   Main themes of the Gita - its essence: original verses with transliteration along with English translation.
21)   From plato to the Gita : similarity between ancient Greek and Indian philosophies.

Dr. Prakash Joshi was honoured with the prestigious J.P. International Award for special and outstanding contribution in the field of education in December, 2024.